THE STORY OF SCIENCE
POWER, PROOF AND PASSION

MICHAEL MOSLEY AND JOHN LYNCH

Left:
Human DNA represented as a series of coloured bands. The Human Genome Project, an international effort to map the genetic code of our species, is one of the most ambitious scientific projects ever undertaken.

Reader's Digest

Reader's Digest Association (Canada) ULC
•Montreal•

The Story of Science: Power, Proof and Passion
Michael Mosley and John Lynch

A Reader's Digest book
This edition published by Reader's Digest Association (Canada) ULC
by arrangement with Mitchell Beazley, Octopus Publishing Group, London.

First published in Great Britain in 2010 by Mitchell Beazley, an imprint of Octopus
Publishing Group Limited, Endeavour House, 189 Shaftesbury Avenue, London, WC2H 8JG
www.octopusbooks.co.uk

An Hachette UK Company
www.hachette.co.uk

ISBN: 978-1-55475-035-1

Address any comments about The Story of Science to: The Book Editor, Reader's Digest Association
(Canada) ULC, 1100 René-Lévesque Blvd. West, Montreal, QC H3B 5H5

To order copies of The Story of Science call 1-800-465-0780 and to discover more fascinating Reader's
Digest products, please visit our website at www.readersdigest.ca

Commissioning Editor: Peter Taylor
Art Director: Pene Parker
Deputy Art Director: Yasia Williams-Leedham
Designer: Mark Kan
Project Editor: Georgina Atsiaris
Copy Editor: Hayley Birch
Proofreader: Jo Murray
Indexer: John Noble
Picture Researcher: Jenny Veall
Production: Peter Hunt and David Hearn

Typeset in Scala and Glypha LT
Printed and bound in China

CONTENTS

INTRODUCTION   6

COSMOS: WHAT'S OUT THERE?   16

MATTER: WHAT IS THE WORLD MADE OF?   56

LIFE: HOW DID WE GET HERE?   102

POWER: CAN WE HAVE UNLIMITED POWER?   144

BODY: WHAT IS THE SECRET OF LIFE?   186

MIND: WHO ARE WE?   230

FURTHER READING   278
INDEX   280
PICTURE CREDITS   286
ACKNOWLEDGEMENTS   288

INTRODUCTION

PALE BLUE DOT

On 14 February 1990, Valentine's Day, the space probe Voyager 1 had reached a distance of some six billion kilometres from the Earth, speeding away from us on its epic journey past the planets and into outer space. The spacecraft had just enough of its precious fuel left to undertake one more special manoeuvre, and on that day the mission controllers gave the instruction. The legendary astronomer Carl Sagan had persuaded them to turn Voyager around, to face back towards its distant home for one last time. Travelling at the speed of light, the signal took 6 hours to reach the spacecraft, but when it did, Voyager responded dutifully. As it turned, laid out before its tiny camera – the camera that over a 13-year mission had faithfully captured the most breathtaking and inspiring images of worlds we never knew could be so strange – was the entire Solar System. Very slowly, one by one, Voyager took a last photograph of every planet it could see, and over the succeeding three months transmitted them back to Earth. The result includes one of the most powerful images of all time: the planet Earth, pitifully small, almost indistinguishable from the thousands of points of light of the stars behind it, appears as a tiny dot of pale blue, less than a single pixel wide, caught in a beam of reflected sunlight bouncing off the shiny surface of the spacecraft. It is a humbling and deeply inspiring picture – all humanity, all our achievements, our futures, all our hopes and dreams, captured in this pinpoint of light.

Left:
A Titan-Centaur rocket soars into the Florida skies on 5 September 1977, breaking the sound barrier as it launches the Voyager 1 space probe on its journey to the edge of the solar system.

But the image of the pale blue dot represents the pinnacle of something else that is special: the knowledge that enabled it to be captured, there, then, at that moment in history. Voyager itself is a product of two millennia of scientific achievement. The material chemistry that built its foil covered skeleton; the mastery of energy that thrust it into space atop a controlled explosion of rocket fuel; the mathematics that understood the opportunity to use a unique alignment of the Solar System to accelerate the probe in a slingshot from one planet to the next; and the quantum physics that allowed its electronics to send back its precious observations of new worlds. Also on board the little spacecraft is a very special cargo, placed there just in case, sometime in the far distant future, it should encounter an alien intelligence. It is a special kind of phonographic record, a gold-plated copper disc containing photographs that summarize our hard-won scientific knowledge – including chemical and mathematical definitions, anatomy and geology. For good measure there are scenes of life on Earth and even sounds such as greetings in 55 different languages, a performance of Beethoven's 5th symphony and a blast of Chuck Berry playing "Johnny B Goode".

SCIENCE NOW

Today, Voyager continues to travel, now some 17 billion kilometres from home, pushing at the edge of interstellar space and carrying with it human scientific achievement in microcosm. Its remarkable mission is the product of questions that have been asked by every human since time immemorial – the great questions: who are we; where did we come from; what are we made of; what is out there? The story of how humanity has sought to answer those questions is the story of science.

In telling that story we will unfold how the modern world was built. For science is embedded in our lives so completely that today we barely notice its presence. Our mobile communication networks depend on orbital mechanics that enables satellites to be positioned in the sky, the chemistry of the rocket fuel that launched them, and the materials that make up the plastics and silicon chips of our computers, phones and batteries. Modern medicine relies not only on intimate knowledge of the biochemistry of each of our cells, but also on an equally deep level of understanding of the atomic structure of matter, in order to scan our organs and bones, to diagnose disease. Our access to the energy that fuels our busy lives depends on our understanding of the geology of the inner Earth and the laws of thermodynamics. Our ability to farm our land and feed our people depends on biologists' capacity to manipulate the process of evolution in the animals and plants that live alongside us. Nothing that we do today is untouched by science, and if we can understand better how that came to be, then we will be better equipped to respond to an uncertain future.

"The history of science is often told as a series of great breakthroughs, revolutions and moments of genius from scientific heroes. But there is always a before, an after and a historical context."

The history of science is often told as a series of great breakthroughs, revolutions and moments of genius from scientific heroes. But in reality there is always a before, an after and a historical context. For science does not happen in a vacuum; it is not set apart in an ivory tower. Science has always been a part of the world within which it is practised, and that world is subject to all the usual complexities of politics, personality, power, passion and profit. So as this story unfolds we will meet characters who worked within the limits of the political and religious climates they knew and were subject to the same pressures as the people who lived alongside them. Only by understanding their world can we understand why the extraordinary advances of science took place when and where they did.

RIGHT PLACE, RIGHT TIME

What is often seen in the history of science is that discoveries emerge from different people at or around the same time. Charles Darwin developed his theory of evolution by natural selection over a number of years in the mid 1800s. Meanwhile, another man, Alfred Russel Wallace, quite independently developed a theory that was in many

Left above:
The cover of Voyager 1's "golden record" offers instructions on how to build a machine capable of playing it. The disc contains sounds and images of distant humanity, alongside a map to locate the probe's origin.

Left:
A grainy image of Earth from more than six billion kilometres away represents one of the pinnacles of human achievement, as well as a salutary reminder of our true place in the cosmos.

ways remarkably similar. Why? Well, the idea that evolution was something that might explain the diversity of the natural world was already being much talked about; both Darwin and Wallace were part of a world eager for travel and exploration, and both had seen things that puzzled them on their voyages; both had read a book by Thomas Malthus that explained how populations are kept in check by famine and disease. But above all, both were part of a historical climate, a society that was driven by overt competition. Victorian life was gripped by the concept of progress, and throughout all layers of society could be felt the consequences of success or of failure to adapt to the rapidly changing commercial and industrial environment. It was in the context of this combination of factors that each of them was inspired to conclude that the driving force behind evolution might be the pressure of natural selection.

It is not just historical events that provide the framework within which scientific understanding has advanced. Technological inventions and discoveries have been critical to the story of science, both directly and indirectly. In the early 15th century the invention of the printing press (more accurately printing with moveable type), attributed to Johann Gutenberg in Germany, resulted in a cascade of consequences for science. The effects of this single event rippled out across the known world, and on over centuries of time, launching the first information revolution. Before the printing press, knowledge was effectively rationed because of the high costs of making books, which had to be copied by hand. At the start of the 1400s, an educated person might have owned a handful of books, if that. After the invention of printing, it was possible to have a library; a collection of books on different subjects – books that did not necessarily all agree with each other. Printed books were used to carry the latest thinking on all subjects – scientific, literary and religious – encouraging the idea of questioning traditional authority. But there is another aspect to the invention of the printing press that can be easily overlooked. Book reading was now a private activity; it did not have to take place in church and it did not have to be supervised. It was one of the many changes that helped to create the more individual, questioning minds that came to make scientific achievements.

More directly, the availability of new technology has frequently resulted in leaps forward in areas of science in which it was suddenly possible to measure and observe things that hitherto would have been unthinkable. The most obvious examples are the invention of the telescope and the microscope, which transformed the understanding of the cosmos at one end of the scale, and the workings of the living cell at the other. Yet often technological invention, and the scientific advances that followed, emerged for very unscientific reasons, such as in the case of the steam engine, which came about as a response to commercial demand – it was the work of practical engineers trying simply to make some money. But once the steam engine was in existence, it became an object of study in itself, as scientists tried to understand the principles of power and energy that enabled it to work. The result of this was the discovery of the fundamental laws of physics that underpin the nature of the Universe.

Above:
The success of tobacco helped to fuel the worldwide search for new species of plants to exploit. The knowledge uncovered changed our view of the origins of life itself.

FOLLOW THE MONEY

As in most walks of life, financial pressures have played a significant part in shaping the progress of science. The story of Galileo's use of the telescope to study the heavens was largely driven by money. When he first heard rumours of the miraculous new invention, the spyglass, the reason he leapt so enthusiastically into action was his difficult financial situation – he was at the time a middle-aged professor of mathematics with limited prospects and badly needed to improve his status and finances. News of the invention of the telescope must have seemed almost heaven sent, an opportunity to impress a new patron amongst the wealthy families of 17th century Italy. He was, of course, utterly unaware of how his brilliant use of the device would come to change the science of the Cosmos.

On a rather grander scale, when explorers and collectors set off on botanical expeditions into the unknown during the course of the 17th and 18th centuries, at least part of their motivation was to find new species of plant that they could exploit. Early adventurers had shown the fabulous wealth that could be obtained from the discovery and sale to the Old World of plants like tobacco. This search for botanical gold led to the unfolding of new knowledge about life across the planet, and fuelled a new understanding of where we, as animals, came from.

"The availability of new technology has frequently resulted in leaps forward in areas of science in which it was suddenly possible to measure and observe things that hitherto would have been unthinkable."

CHARACTER COUNTS

All scientific discoveries owe much to the particular historical context that their discoverers find themselves in. Some, however, also seem to depend critically on the discoverer's character. A good example of this is Johannes Kepler. At the beginning of the 17th century, working alone in Prague, Kepler discovered three laws of planetary motion that would in time transform our view of the Solar System. He would not have done what he did without the political and religious changes brought by the Reformation, which undermined belief in established authority and drove him from his home. He needed the financial and political support of the Holy Roman Emperor, Rudolph II – an Emperor obsessed with horoscopes who had the money to pay for stargazers. And he certainly needed data that had been painstakingly collected over many years by his colleague Tycho Brahe. But he also needed a spark of genius and an obsessive personality.

At a time when almost everyone believed the Earth was at the centre of the Universe, Kepler believed passionately that the Sun was the symbol of God and produced a force that drove the planets around itself. To prove this, he realized that he had to describe orbits for the planets that would match what was seen in the night sky. This is where his obsessive nature proved to be so important, because this was complex and unbelievably tedious work. He kept at it for five long years, producing hundreds of pages of calculations before he finally reached a solution. As he later wrote, "If thou dear reader are bored with this wearisome calculation, take pity on me who had to go through it seventy times."

Leonardo da Vinci was a very different sort of character. He too was fascinated by the stars, but he was also fascinated by a great many other things. This, in a way, was his problem, for he was rarely ready or willing to focus long enough to complete what he had started. Along with his painting and his inventions, he produced some of the finest anatomical drawings of the human body ever made, with the intention of creating a textbook of anatomy. The problem was that, like so much else he did, he never finished it. By the time his work was rediscovered, others had done what he set out to do, and he remains forever described as a man who was ahead of his time.

Others who have left their mark on science have shown a much more ruthless streak than either Kepler or da Vinci. Sir Isaac Newton, a strange and obsessive character, took his reputation so seriously that he was prepared to destroy that of anyone who crossed him. Robert Hooke, his intellectual rival for many years, is just one of those who endured his anger and vindictiveness. Newton's malevolent behaviour, however, pales in comparison to the competitive drive of the inventor Thomas Edison. Wanting to demonstrate the superiority of his method of electrical power distribution, Edison quietly supported the development of the electric chair – which was reliant on a competing method – to show that the alternative was lethal, and thus to discredit the work of his rival.

SHEER CHANCE

Scientific advances are the product of the same social pressures and subject to the same human triumphs and failings that beset all other walks of life. But blind chance has also played a significant role throughout the history of science. As Louis Pasteur, the father of germ theory and codeveloper of pasteurization, once famously said, "chance favours the prepared mind". The point is that it is important to be in the right place at the right time but that chance alone is not enough. One of the most famous examples of a serendipitous discovery was that made by Alexander Fleming.

In 1928, Fleming went on holiday, leaving behind a petri dish with bacteria growing on it. Somehow fungal spores being investigated by one of his colleagues in another part of the building made their way into his laboratory and settled on the dish. The weather, it so happened, was balmy and created perfect conditions for both bacteria and fungus, so that when Fleming returned from holiday and picked up the dish he was able to say to himself, "Now I wonder what is killing the bacteria?" Luck was certainly on his side; he was extremely fortunate to discover the antibacterial properties of penicillin. But the point is that he had spent a large number of years looking for something that would kill bacteria, so he swiftly recognized the significance of what he saw. Instead of throwing the dish away, he investigated.

"Each explanation that science has offered of how the world works is very largely a product of its time."

The chemist William Perkin also made a discovery that was utterly unexpected, and demonstrated the flexibility and wit to exploit an opportunity when it came along. In 1856, aged just 18, he was in his parents' house trying to produce a synthetic version of the anti-malarial drug quinine. Instead of a quinine substitute, he found that he had accidentally created a chemical that could dye a bit of cloth an intense purple. Many chemists might have put this discovery aside as being of minor importance, but Perkin realized its commercial potential. Not only did he make his fortune, he also created a new and important branch of industrial chemistry.

As we shall see as the story of science unfolds in this book, over centuries the forces of history, personality, money and technology have all come to bear on the moments in which each of the great scientific advances has occurred. Each explanation that science has offered of how the world works is very largely a product of its time.

Right:
Louis Pasteur's lifesaving research into killing microbes began with his early experiments in the production of a disease-free beer. His work exemplifies his belief that "chance favours the prepared mind".

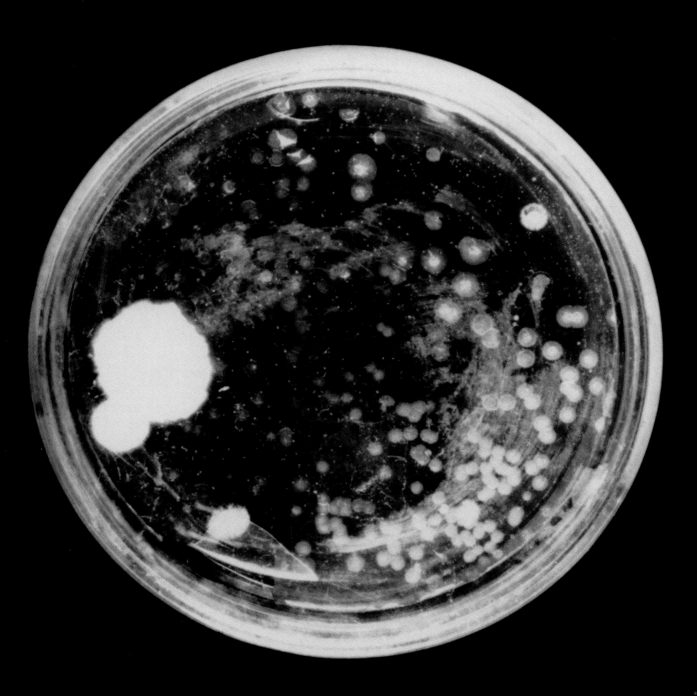

Print of the culture plate which started
the work on Penicillin

UNCOMMON SENSE

But as science has advanced it has also often struggled to find acceptance among the wider population. Part of the problem is that we are used to having opinions about the world and, on the whole, having those opinions respected. Yet when it comes to science, the opinions of the many do not count. Science is not, in that sense, democratic. Many of its truths are counterintuitive, and our everyday experience of the world is largely irrelevant to its progress. When it comes to science, common sense will only take you so far. Can there possibly be an attraction between the Earth and the Sun that operates instantly and across huge tracts of empty space? How plausible is it that whole countries sit on top of great slabs of rock that slowly float across the surface of a molten core? Can we, and every creature on Earth, really be descended from the same single cell? The world around us seems reassuringly solid, so what does it mean when scientists tell us that everything is made of atoms and that atoms themselves consist almost entirely of emptiness?

In daily life, we end up taking these things largely on trust, as for the most part they do not affect us directly. But when there is conflict between evidence and our self interest or our personal observations of the world, problems can arise. Today, the evidence in support of manmade climate change is overwhelming, yet throughout the debate over global warming many people have remained unconvinced that it is a real phenomenon – partly because the science is so complex and the detail so uncertain, and partly because the consequences for our lives are so difficult to accept. Similarly, Darwinian evolution is the most robust of scientific theories, but many people find it easier and more reassuring to believe that something so intricately constructed as life on Earth must have had a designer. It is hard for us to accept that everything around us, from the beauty of the butterfly to the complexity of the human eye, could have evolved simply through chance and the pressure to survive. But the scientific evidence that evolution by natural selection has occurred is unassailable.

The scientific method, which is today practised in all corners of the globe, builds explanations based on evidence, and when new evidence emerges that does not fit with the model, the explanation must change. That is how science moves on. Ultimately, however doubtful the motives, however influential the economic, political and personal forces that have born down on the men and women who practise it, and however unpalatable the conclusions are, the development of science has depended on following where evidence leads. The path that we have followed in search of answers to our big questions has been an epic one, full of rich characters and inspirational ideas, moving in fits and starts, pursuing blind alleys, surmounting great hurdles and frequently taking unexpected turns. It has been one of the most colourful and exciting stories that it is possible to tell.

Left:
A photograph of Alexander Fleming's original culture of the Penicillium fungus. Though serendipity played a great part in the initial discovery, it was the methodical work of other scientists that turned it into a viable treatment.

Cosmos

WHAT'S OUT THERE?

If, on a clear night, you are lucky enough to find a secluded place from where you can gaze up at the night sky and be awed by the enormity of what you see, it can be a humbling and moving experience: a canopy of stars spread above you, inspiring wonder; everything appearing quiet, still, and peaceful. Yet we are actually on a large rock spinning around its own axis at over a thousand kilometres an hour. We and our companion, the Moon, are also travelling in a giant loop around the Sun at about a hundred thousand kilometres an hour and we are held in this endless swirl by a warping of space-time that we call gravity. We are just part of one tiny galaxy that, along with thousands upon thousands of others, is being pulled through space towards a gravitational anomaly in the intergalactic void. These are not intuitively obvious concepts to grasp; indeed they go against all our common-sense experience, because to all intents and purposes we live on a static surface that is daily lit by a rising and setting sun. It is therefore not surprising that it took much time, much toil, and much anguish to arrive at the understanding of the cosmos we have today, and our answer to the question: what's out there? It is a remarkable story and it centres on a time of terrible upheaval some four hundred years ago.

Left: This Hubble Space Telescope image captures details within a barred spiral galaxy similar to our own Milky Way. This independent system containing billions of stars is so far away that its light has taken 70 million years to reach Earth.

THE COURT OF THE EMPEROR

The forces of change that swept across 16th-century Europe were powerful, fast moving, and deadly. The Protestant Reformation, started in 1517 as a protest against corruption in the Church of Rome, brought more than a century of turmoil to the fledgling nations of Europe. Yet through all the violence that ensued one deeply reassuring idea held sway, as it had for millennia: the Earth was, self-evidently, at the centre of everything. Around our secure and fixed world moved the heavens, with their pinpoints of starlight, their wandering planets, and the life and light-bringing Sun and Moon; and all of this had been arranged there for us by God, whether that god was mediated by the pronouncements of the Pope or interpreted literally from a Lutheran Bible. It was an idea in which the power and authority of both Church and State was vested. Yet quietly, throughout the turmoil, building gradually as the century drew to a close, that view was to be revolutionized, and the certainty and authority that it brought was to be shattered.

The roots of that transformation can be uncovered among the characters of the flamboyant court of the Holy Roman Emperor Rudolf II, King of Bohemia, King of Hungary, Archduke of Austria and Moravia, who reigned from 1576 to 1612. He was a melancholic and withdrawn ruler whose primary interests lay in the occult and in learning, rather than in the machinations of politics and government. But above all, Rudolf was a collector. He gathered a menagerie of unusual animals, original artworks, clocks, scientific instruments, and a botanical garden, and had a special wing built at Prague Castle to house his spectacular collection of "curiosities", which included almost everything from a gem-encrusted rhino horn to what he thought were the feathers of a phoenix, a mummified dragon, and the horn of a unicorn.

Rudolf also collected people. His passion for learning attracted some of the best minds of the age to his court, and his tolerance of different religious beliefs created an atmosphere of dialogue and discussion around him. It was to his court in Prague that travelled two of the most significant people in the story of our understanding of the heavens – and one of the oddest couples in the history of science. One was an arrogant Danish aristocrat with the finest assembly of astronomical instruments of the age, and the other was a poor, German mathematician on the run from religious turmoil.

Below: The Basilica of St. Vitus towers over an enormous maze of buildings that makes up Prague Castle.

THE IRASCIBLE DANE

Tycho Brahe
1546–1601

His false nose and apparently enlarged left eye can be seen in the portrait.

The year 1560 was historically unremarkable. But on 23 August there occurred an event that made a deep impression on the first half of our odd couple – a total eclipse of the Sun. In Copenhagen it was seen only as a partial eclipse, but the fact that the event had been predicted on the basis of tables of observations of the movement of the stars and Moon seemed of great significance to a young Danish aristocrat boy, Tycho Brahe, and stimulated a passion in him for star gazing. In his mid-teens he studied law, but also began to buy up astronomical instruments and books, beginning a lifetime of making observations of the night sky. What struck the young Tycho was the variability of the observations recorded in star tables that had been passed down from the ancients. Aged just 17, he had written: "What is needed is a long term project with the aim of mapping the heavens conducted from a single location over a period of several years." That was what he set out to do, and by the end of his life he had provided the fundamental evidence that was needed to answer the question of what is out there.

Tycho moved on through several of the great universities of Europe – Wittenberg, Rostock, Basle, and Augsburg – studying astrology, alchemy, and medicine, and amassing an extraordinary collection of astronomical instruments. As a student, however, an event occurred that literally marked him out for life. Tycho got into a fierce argument with another scholar while at a dance. We do not know for certain what the argument was about, but it ended in a duel, fought in the dark, the result of which was that he lost a portion of the bridge of his nose. For the rest of his life he wore a special cover made of metal – it is said he wore a copper alloy for everyday use and gold or silver as formal wear.

"By the end of his life he had provided the fundamental evidence that was needed to answer the question of what is out there."

Right: One of Tycho's most famous achievements was an enormous model of the celestial sphere, 180cm (6ft) in diameter. Sadly, it was destroyed in a fire at Copenhagen University in 1728.

The silver nose underlined Tycho's reputation as a high-handed and irascible aristocrat, but his reputation as an observer of the heavens was not established until a few years later with the appearance of a remarkable new star in the night sky. By then he was back in Denmark, and on the evening of 11 November 1572 he was walking back to his house from his alchemical laboratory when he noticed a very bright new object in the constellation Cassiopeia. Tycho had spotted what we now know as a supernova – the explosive death of an ageing star. Indeed, it was he who coined the term "nova" when he published a book on the discovery, *De Nova Stella*. As he observed it assiduously over the following months SN1572, as it is known today, gradually dimmed, but what he saw was quietly shocking for the model of the Universe that he and all around him held to at the time. As he built up a dossier of observations it became apparent that it was indeed a star, not a planet, nor a comet. Here was something new appearing in what everyone believed was the "crystal sphere" of the fixed stars that had been created, perfectly formed, by God. For a new star suddenly to emerge in the heavens was a chilling idea.

Below: Newborn stars illuminate huge regions of pinkish star-forming gas in this Hubble Space Telescope image of an "irregular" galaxy.

DUTCH GLASSES

War continued to be a backdrop to science throughout the 16th and 17th centuries. Late in 1608, there was a peace conference in the Netherlands, which ushered in what became known as the "Twelve Years Truce" in the Eighty Years War, in which the Protestant Dutch fought to shake off the rule of Spain. At the conference, Prince Maurice of Nassau was shown a "spyglass" by Hans Lippershey, a spectacle maker from Middelburg. Spectacles had been around for hundreds of years by then, and spectacle makers had mastered the tricks of shaping concave lenses to correct short-sightedness and convex ones to assist reading. What Lippershey did was to put one of each together to produce a rudimentary telescope. That same year he had tried to get a patent on his invention, but it was refused because two other claimants came forward at roughly the same time. Why the idea suddenly emerged then in Holland is unclear, but its potential on the battlefield was immediately obvious, and the presence of many diplomats at the demonstration meant that news of it spread rapidly across Europe as emissaries travelled home with the glad tidings of peace. By early 1609, small spyglasses could be bought at the Pont Neuf in Paris, and by the summer they had reached Italy.

> *"The device was very simple: a tube with a convex lens at one end and a concave one at the other, producing a magnification of about three times."*

Left: Galileo Galilei 1564–1642. This portrait dates from 1636, during the period of Galileo's house imprisonment.

The device was very simple: a tube with a convex lens at one end and a concave one at the other, producing a magnification of about three times. When news of the spyglass reached a professor of mathematics in Padua, he thought he could probably make one too. The mathematician was Galileo Galilei, a combative, bombastic man determined to succeed and ever on the lookout for a chance to improve his position amongst people who mattered. Here was an opportunity to impress his potential patrons, and so he planned to build a spyglass for the Doge of Venice, under whose authority he lived and worked in Padua. In the summer of 1609 came news that a Dutchman was about to present a telescope to the Doge, and in an extraordinary piece of practical craftsmanship Galileo had his own version built in 24 hours and then improved it to eight times magnification. This he managed to present to senators of the republic, ahead of his competitor, in August of that year. The result was a sensation. Galileo wrote that several times he and Venetian gentlemen of influence climbed the highest bell towers in the city to view distant objects greatly magnified. His submission to the Doge emphasized the advantage it would give the maritime state in its constant need for defence against attack from the Ottoman Turks, "allowing us at sea to discover at a much greater distance than usual the hulls and sails of the enemy, so that for two hours and more we can detect him before he detects us". This was how the world worked in the time of Galileo: a significant gift was offered to a patron, together with a gentle hint at the need for personal gain. The Doge responded generously, offering to double Galileo's salary and confirm his position in Padua for life. However, the small print was less attractive: life meant exactly that; he would not be allowed to move, and the salary would never change. So Galileo set his sights elsewhere.

Right: A pair of Galileo's early telescopes are now displayed side by side at the Museum of the History of Science in Florence.

PERFECT VISION

Venice at the turn of the 16th century had passed the peak of its Renaissance glory. The great Venetian rowing galleys, which had dominated Mediterranean trade and numbered some 3,000 at their peak, had been overtaken by the sea-going ships that now plied the oceans to the west. The fall of the Byzantine city of Constantinople to the Ottomans had resulted in a reawakening of classical knowledge as both people and texts crossed the Aegean to safety in Italy, bringing understanding that helped to fuel the flowering of the cultural and humanist explosion of the Renaissance. At the same time the need to find other routes to the Far East, to bypass the Ottoman power, opened up the drive for exploration and the discovery of the New World – the Americas. The result was that economic power was shifting away from Italy towards northern Europe. But Venice was still a cultural powerhouse, and among its crafts was glass. As far back as the 1200s, Venetians had brought sodium ash back from the Middle East and this was one of the key ingredients of the remarkably pure crystal glass that their glassmakers produced on the island of Murano in the Venice lagoon. Indeed, the Murano craftsmen were prohibited from leaving Venice, in order to try to keep their special skills secret. It was this glass that helped Galileo to make dramatic improvements to his telescope over the autumn of 1609; by the end of the year he had achieved a magnification of 20 times.

Galileo had also begun turning the telescope to the heavens, starting with a study of the Moon. What he saw, and revealed in a series of eight drawings across its different phases, was startling. Instead of the perfect smooth sphere defined by Aristotle's cosmology, he saw rugged mountains and jagged edges to its circumference and shadow-line. He went on to look at other heavenly bodies. In January of 1610, he saw what seemed to be three small stars in a line across Jupiter; three nights later one had disappeared. Three nights after that it was back, in a different position, and he also

Right: Spaceprobes have revealed the four tiny points of light that Galileo discovered orbiting Jupiter as worlds in their own right. Here the volcanic moon Io (left) and icy Europa hang in front of Jupiter's own turbulent cloudscape.

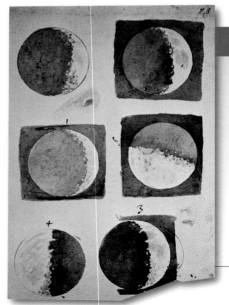

GALILEO'S MOON OBSERVATIONS

These pioneering observations were first made using the telescopes that Galileo had built in 1609. Galileo saw that the terminator (the line between the Moon's night and day sides) was sometimes irregular (top) and sometimes smooth (bottom). He deduced that the irregularities were due to mountains on the Moon, the first time Earth-like objects had been discovered in the heavens. This challenged the existing world view that said the heavens were perfect and unchanging. This page is from a 1653 edition of *Sidereus Nuncius* (March 1610).

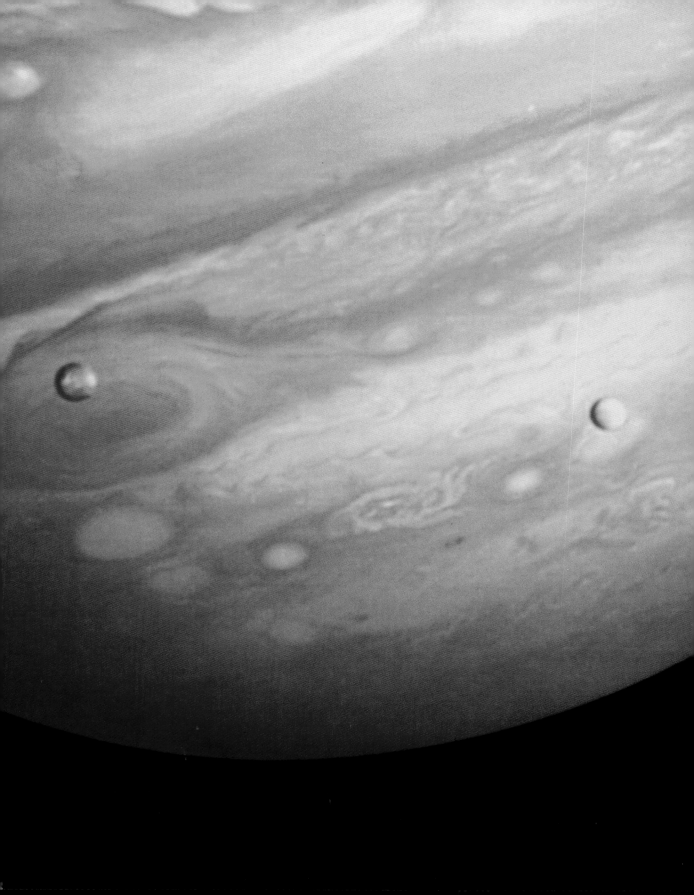

saw a fourth. He realized that these "stars" must be orbiting Jupiter. He had discovered the planet's moons. Galileo was already favourably disposed to the Copernican view of the cosmos, but this was powerful evidence, albeit indirect, in its support. For if Jupiter had moons that orbited around it, then the Earth's special position at the centre of everything was no longer so special. Galileo decided to publish his discoveries, and wrote them up in a short, very accessible book written in Italian rather than Latin and including pictures. He called it the *Siderius Nuncius* – the "Starry Messenger". Ever the opportunist, he dedicated it to the Grand Duke Cosimo II di Medici, and proposed naming the four new satellites of Jupiter after the four Medici brothers. As a tactic, it worked. By the summer of 1610 he had been appointed as court mathematician and philosopher to the Medicis in Florence, at a hugely increased salary. His future status and wealth were secured.

"If Jupiter had moons that orbited around it, then the Earth's special position at the centre of everything was no longer so special."

THE TRIAL

The story of Galileo's later life – his clash with the Church, his trial for heresy, and his imprisonment – has stood out as a turning point in the history of science, but the reality is not quite as simple as it is often portrayed. In the end, it was less about heresy than about authority. For what Galileo began to do was to stray into territory that the Church believed should be firmly under its control. The observations Galileo made with his telescope gave him more confidence in Copernican theory. Not only did he see the moons of Jupiter and the uneven lunar surface, but also he observed the complete phases of Venus – its changing illumination as it orbits the Sun, falling into and out of shadow, as it passes in front of the Earth. His findings, at first, were accepted for what they were, simply observations with no consequent interpretation, and Galileo had a successful audience with the Pope in 1611, with much glory being attached to his patrons in Florence as a result. But thereafter things gradually began to go wrong. He antagonized an influential Jesuit astronomer by over-claiming the discovery of sunspots, and he became more public in his support for Copernicanism. In a letter to the Grand Duchess of Tuscany, in response to her concerned questioning as to whether the theory went against biblical teaching, Galileo famously wrote: "One must not begin with the authority of scriptural passages, but with sensory experience, and necessary demonstrations." This is one of the earliest expressions of what we now think of as the scientific approach – evidence first, deductions follow. But it also reveals that Galileo was beginning to say that his world of scientific study might have equal status to revelation. This was dangerous ground. In 1616, after a Papal commission concluded that Copernican theory was indeed heretical, Galileo was instructed not to "hold to or defend" the heliocentric view.

Left: According to a popular tale, at the end of his trial a humiliated Galileo forced to recant his beliefs, still muttered under his breath "nevertheless, it [the Earth] moves!" Sadly, the story first appears a century after the trial itself, and there is no evidence that it is true.

DIALOGO
D I
GALILEO GALILEI LINCEO
MATEMATICO SOPRAORDINARIO
DELLO STVDIO DI PISA.
E Filofofo, e Matematico primario del
SERENISSIMO
GR.DVCA DI TOSCANA.
Doue ne i congreffi di quattro giornate fi difcorre
fopra i due
MASSIMI SISTEMI DEL MONDO
TOLEMAICO, E COPERNICANO;
Proponendo indeterminatamente le ragioni Filofofiche, e Naturali
tanto per l'vna , quanto per l'altra parte .

CON PRI   VILEGI.

IN FIORENZA, Per Gio:Batifta Landini MDCXXXII.
CON LICENZA DE' SVPERIORI.

Above: Galileo published his *Dialogue* in Italian rather than Latin, ensuring his ideas would reach the largest possible audience.

There it lay, but once again Galileo indulged in a public spat with Jesuit astronomers, in a critique he wrote of their theories on comets. And in a book that followed he underlined his philosophy that the laws of the Universe "are written in the language of mathematics". Yet again, an audience with the new Pope Urban VIII, who had been a personal friend, went well, and Galileo was encouraged to write a book that compared the two world views – the Ptolemaic and the Copernican – providing, of course, that he did not claim that the Copernican model was true. The result in 1632 was the *Dialogue Concerning the Two Chief World Systems*. In it he set out the arguments for and against each model in the form of a debate between three characters. It is generally agreed that he went too far, allowing himself to be identified with the defender of Copernicus, and putting the case for Ptolemy into the mouth of a character he called "Simplicio" – the simpleton. The Inquisition saw its chance and the trial of Galileo, for breaking the injunction not to defend Copernicanism, was engineered. Famously, he confessed guilt and recanted; the *Dialogue* was banned. Galileo was declared a heretic and placed under house arrest for the remainder of his life. But most importantly, Vatican authority over knowledge was restated: truth lay in the teaching of the Church.

FATHER OF SCIENCE

As is ever the way, the long-term outcome of suppression was not what the factions inside the Church might have hoped for. The book was smuggled out of Italy and the intellectual community across Europe read it avidly. Within a generation, few educated people still believed that the Sun travelled round the Earth, although nothing that Galileo had done could be regarded as incontrovertible proof. Indeed, that had to wait for another century, when instruments were good enough to observe stellar parallax – the shift in the relative position of the stars and the Earth, as we orbit the Sun. But after Galileo, and after Kepler, the question became less to do with how the planets move as with what holds them in place.

Arguably, Galileo's greatest work was in an entirely different realm of "natural philosophy". He had spent his life doing mathematical and experimental work in the realm of physics, fluid science, and mechanics. He laid out mathematical laws of the movement of projectiles and the acceleration of falling bodies; he famously proved that objects of different sizes and weights would fall at the same speed (although stories that tell of him dropping cannon balls and feathers off the Leaning Tower of Pisa are nothing more than stories); and he demonstrated that it was the density of an object rather than its shape that determined whether it would float.

Galileo died in January 1642 at the age of 77. The family villa where he was incarcerated, at Arcetri in the hills above Florence, still carries an air of calm isolation and solitude. For his last decade the old man, gradually growing blind and more infirm, spent his time there pulling together the results of his lifetime's work. Above all, everything he had done was based on observation and experiment – gathering evidence and building arguments from what can be seen and reproduced is at the heart of a modern scientist's work. That is perhaps why Galileo has been called the "father of science".

THE HIGH SEAS

The opening of the sea routes to the Americas by Christopher Columbus in 1492, and the heroic ocean voyages that followed, transformed the world. The 16th century was the era of global circumnavigation, with the first expedition around the world in 1519 heralding an unprecedented territorial expansion of the maritime European countries. The wealth in gold and silver that was extracted from the New World and found its way back to Spain, Portugal, the Netherlands, and Britain was vast. Silver from the Americas alone accounted for a fifth of Spain's national budget in the late 16th century. This new global trade and ocean exploration brought with it a far greater urgency to solve a problem that had been recognized for centuries: how to arrive at an accurate measurement of longitude, a position east or west around the globe. Throughout the early age of expansion there was real economic pressure to find a solution in the measurement of stars, Moon and planets. The Spanish and the Dutch governments both offered prizes for a solution; the French founded the Académie Royale des Sciences, charged with improving navigation, and in England the Royal Observatory was founded in 1675 at Greenwich. Astronomy became the most important scientific endeavour of the time, and the topic of everyday conversation amongst the intelligentsia of Europe.

Left: When Columbus's fleet landed in the Bahamas in 1492, he initially believed he had arrived in the Far East. The opening up of East-West trade routes in the following century only added to the need for an accurate method of determining longitude.

LATITUDE AND LONGITUDE

Early astronomical instruments, such as the cross-staff, the astrolabe, or the quadrant, were easily sufficient for determining a sailor's latitude (position north or south of the equator), by measuring the angle between the horizon and the Sun or known stars. Longitude is very different. Because the planet rotates once every 24 hours, the solution ultimately lay in the ability to know the time at your home port when it was, say, noon at your position on the ocean. The time difference will tell you how far round the globe you are. But a reliable, accurate, ship-board clock (the chronometer, famously perfected by a Yorkshire carpenter, John Harrison) was only available from the late 1700s, and then barely affordable until the mid 19th century. An alternative method was known as "lunar distance", and depended on knowing how the position of the Moon would appear relative to the stars at different points around the globe. This required the compilation of lunar distance tables, based on thousands of observations. Until recently ships always carried lunar distance tables with them, in case of the failure of their chronometers.

THE BET

By the late 16th century, global trade had brought great wealth to the City of London. The repeated attacks of plague had passed, and the Great Fire of 1666 had triggered a huge programme of building and renewal. Everywhere in the capital, coffee houses had sprung up and people gathered in them to discuss the matters of the day; to do deals, to argue and debate rival theories of the world, to listen to stories of distant lands, and above all to exchange information. In 1684, three men regularly met at Jonathan's and Garraway's coffee houses and it may well have been at one of these two establishments that they agreed a bet. Christopher Wren, architect of the rebuilding of London, Edmond Halley, yet to become linked with the comet that today bears his name, and Robert Hooke, the curator of experiments at England's national academy of science, the Royal Society in London, were debating the question of what actually held the planets in their elliptical orbits around the Sun. At that time the favourite explanation was magnetism. During the conversation they challenged each other, for the sum of two pounds – or about 250 cups of coffee – to demonstrate that the force, whatever it was, obeyed the inverse square law, namely that the force gets weaker in proportion to the distance, squared, of the planet from the Sun. This was an idea that had been discussed for a number of years, but mathematically, it was a very difficult thing to prove.

"They debated what held the planets in their orbits. The favourite explanation was magnetism."

Below: Coffee houses were popular meeting places in the 17th and 18th centuries, and the scene of many scientific innovations.

GRAVITY

Later that year, Halley visited Cambridge and called to see Isaac Newton, the Lucasian Professor of Mathematics at the University, and a man with a formidable scientific reputation, especially for his work on optics. Halley put the question of the inverse square law to Newton and was astonished when Newton told him that he had already proved it – but he just could not find the paperwork there and then. The encounter prompted Newton, encouraged by Halley, to go back over his work of some 20 years previous and compile it into a book. By rights, the publication should have been funded by the Royal Society, of which both Halley and Newton were influential members, but the Society had recently lost substantial sums on publishing a natural history of fish, by one Francis Willoughby, that had failed to sell. So Halley stepped in and paid for the printing himself (and was repaid in copies of the fish book). It is fortunate that he did so, for Newton's work, published in 1687, stands as one of the greatest contributions to science of all time. It was called *Philosophiae Naturalis Principia Mathematica* – the *Principia* for short.

The history of science is often told as stories of individual brilliant minds, flashes of inspiration, and men leaping out of baths shouting "eureka". The reality is very different: ideas emerge into the zeitgeist and become talked about; technological advances make things possible to see or understand; historical events open up opportunities or pressures for change. There is always a context. Isaac Newton and the advances he made were part of all that, but he is truly one of the few greats to whom the word "genius" can perhaps be applied. Born in 1643 at Woolsthorpe in Lincolnshire, Newton was a complex, insular man. His childhood and youth had been miserable. He once even threatened to burn down his house with his mother and stepfather inside. He was educated away from home, and then brought back at the age of 16, when his stepfather died, to run the farm. He was a terrible farmer and so was sent to Cambridge University as a "subsizar", a position where he had to act as a servant to other students to earn his education. Then, to escape the threat of plague, he returned to Lincolnshire in 1665, and it was there, over just two years, that he did the science and mathematics for which he would become most famous. He went back to Cambridge to become a fellow of Trinity College in 1667 and matured into a secretive, reclusive, and obsessive man – but an intellectual giant. It is always easy to think of Newton and his achievements as a shining example of scientific thinking at work – a beacon of the Enlightenment and so on – yet he had unusual religious beliefs that made him a near heretic and saw his principal aim as to explain the mind of God. What's more, the bulk of his life's study was spent in pursuit of alchemy. In reality, he sits uneasily between the occult and mysticism on one side and the beginnings of true science on the other.

By the time Edmond Halley visited him in 1684, Newton had already shown that the colours of the rainbow could be passed through a prism to recombine as white light, thus proving the nature of the spectrum. He had already invented a new form of mathematics, which we now know as calculus, which enabled the analysis of movement. And he had, as he told Halley, already worked out the nature of the force

Isaac Newton
1643–1727

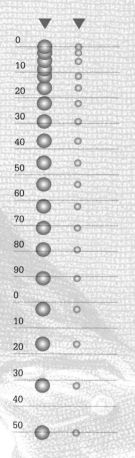

Above: The law of falling bodies, famously identified by Galileo, demonstrates that, air resistance aside, two bodies of unequal weight will accelerate downwards at the same rate in Earth's gravitational field, and hit the ground after the same time.

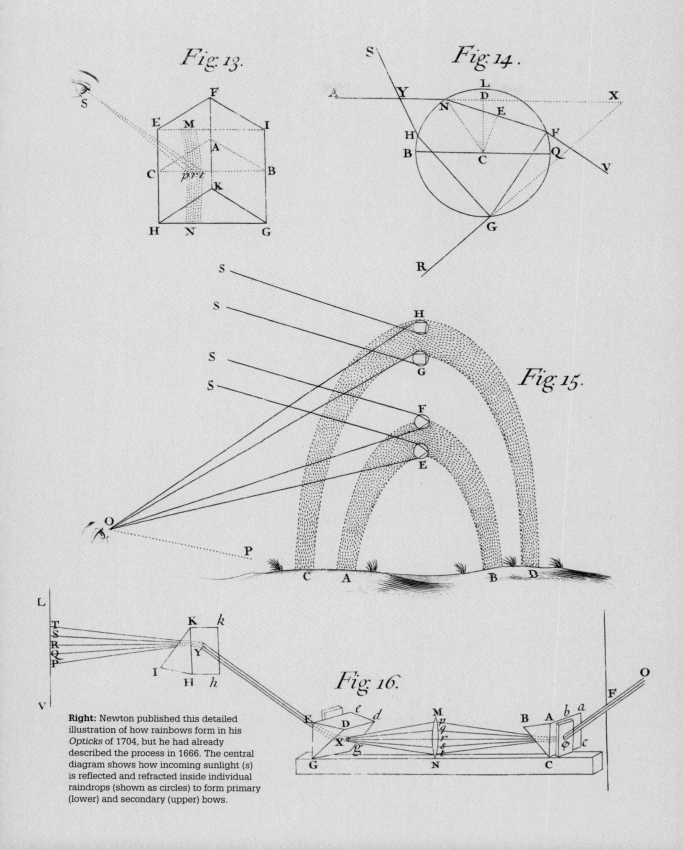

Right: Newton published this detailed illustration of how rainbows form in his *Opticks* of 1704, but he had already described the process in 1666. The central diagram shows how incoming sunlight (s) is reflected and refracted inside individual raindrops (shown as circles) to form primary (lower) and secondary (upper) bows.

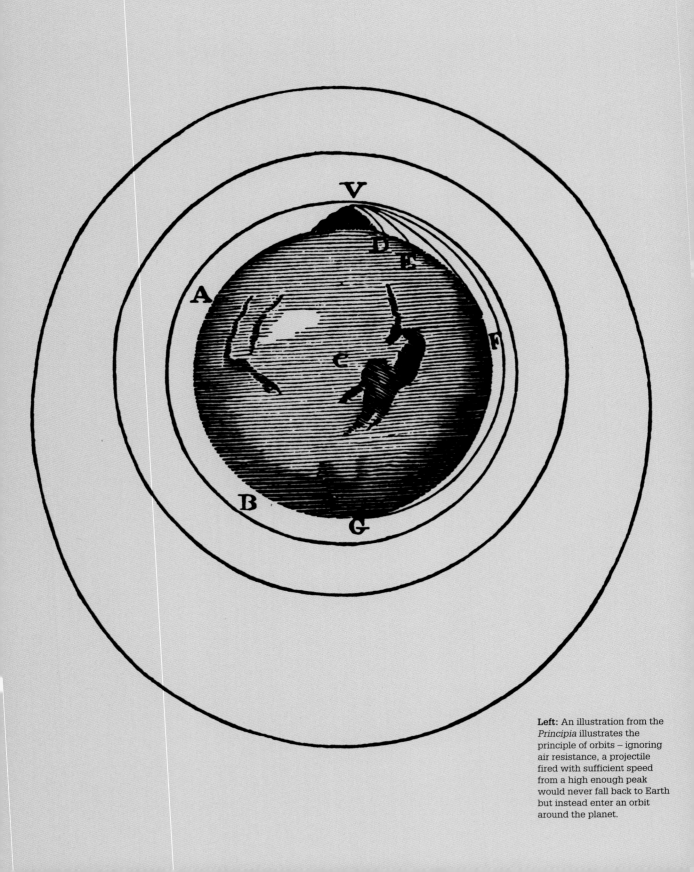

Left: An illustration from the *Principia* illustrates the principle of orbits – ignoring air resistance, a projectile fired with sufficient speed from a high enough peak would never fall back to Earth but instead enter an orbit around the planet.

PHILOSOPHIÆ

NATURALIS

PRINCIPIA

MATHEMATICA

Autore *JS. NEWTON*, *Trin. Coll. Cantab. Soc.* Mathefeos
Profeffore *Lucafiano*, & Societatis Regalis Sodali.

IMPRIMATUR
S. P E P Y S, *Reg. Soc.* P R Æ S E S.
Julii 5. 1686.

L O N D I N I,

Juffu *Societatis Regiæ* ac Typis *Jofephi Streater*. Proftat apud
plures Bibliopolas. *Anno* MDCLXXXVII.

Above: Isaac Newton's
Principia was finally
published in three volumes
in 1687. Only around 300
copies were printed in the
first edition and about 1,000
in the second.

that holds the planets in orbit around the Sun. He had done the fiendishly complicated maths to show that the force would indeed result in elliptical orbits, fitting in with Kepler's laws of planetary motion. And he had combined this mathematical description of the orbits with the results of Galileo's experiments on the movement of projectiles and falling bodies, to explain why the planets stay in orbit.

UNIVERSAL LAWS

The *Principia* defined the nature of the force of gravity and set out laws of motion for everything from a cannon ball, to the Moon, to a planet orbiting the Sun – or to an apple falling from a tree. The famous story of his watching the apple fall as a young man, and realizing the same force could be extended to the Moon, is almost certainly a complete fiction – a story made up by Newton himself in later life to ensure that he retained credit for the discovery, and to immortalize himself. For there was a side of Newton's character that was deeply unpleasant. He engaged in lifelong and vindictive disputes with the first Astronomer Royal, John Flamsteed, and with Robert Hooke of the Royal Society, who himself came close to or even matched Newton's achievements in optics and defining gravity. Newton even carried the dispute beyond the grave. After Hooke's death, Newton, by then president of the Royal Society, allowed Hooke's portrait to be "mislaid" during a move of premises, so that the image of his arch-rival was lost to posterity.

The most powerful thing about Newton's laws is their universal nature. For Newton, his work brought him closer to his goal of seeing the mind of God, but for the history of science it marks a rare turning point. He had shown that the laws of physics can be applied to everything. Today, our modern world depends on our understanding of those laws, for example, in the acceleration of the rockets that launch communications satellites into space and the geostationary orbits such

NEWTON'S LAWS OF MOTION

FIRST LAW An object will stay at rest or continue in motion unless another force acts upon it. For instance, a ball will keep rolling unless friction slows it down.

SECOND LAW The force acting on an object is defined as its mass multiplied by its acceleration,

$$F=ma.$$

THIRD LAW For every action there is an equal and opposite reaction. For example, the recoil of a gun as a bullet speeds away.

To show how the force of gravity would hold a planet in orbit, Isaac Newton devised a thought experiment. He imagined a cannon on top of a high mountain, far above the atmosphere. He thought: if a ball is fired from the cannon slowly then gravity will pull it down to Earth. However, if the ball is fired with enough force it will escape gravity and disappear off into space. But if the velocity is just right the ball will keep travelling right around the Earth, held by gravity in a giant orbit, just like the Moon.

satellites adopt around the Earth, enabling our dependence on modern communications to continue. In the century that followed the publication of Newton's work, the idea of universal laws took off, and became applied in many other fields: economics, the living body, even the mind. The word "Newtonian" itself came to represent the notion of perfectly ordered and mathematically defined theories, those worthy of being taken seriously.

FOLLOW THE MONEY

The mid 18th century was a glorious time for astronomy. In pursuit of the solution to the longitude problem, money was poured into the study of the stars. The Royal Observatory flourished, and George III – better known today as "Mad King George" – was an eager supporter of science. Telescopes got bigger, and so did the Solar System. King George may have been bitter at the loss of the Colonies in the American War of Independence, but in 1781 there was some recompense when the astronomer William Herschel discovered a new planet, which he named "George's Star" in honour of his patron. The name was not an international hit, however, and by the mid 1800s the world had settled on "Uranus", suggested by a German astronomer, after the father of Cronus (Saturn's equivalent) in Greek mythology. By then, yet another planet had been spotted – Neptune, in 1846 – whose position had been predicted mathematically in a triumphant confirmation of Newton's laws, because of a perceived distortion in the orbit of Uranus due to the gravitational pull of the unseen planet. The 20th century brought yet another discovery, with Pluto completing the set as the ninth planet in 1930. Sadly for Pluto it has recently been downgraded to a "dwarf planet", so the 21st-century Solar System numbers just eight planets.

Each of these great discoveries was dependent on bigger and better telescopes. It meant that astronomy remained a very expensive science and, as so often throughout this story, the astronomers "followed the money". By the turn of the 20th century, the money was in the USA, which was well on the way to becoming the dominant economic power that it is now. The industrial and financial barons of late 19th-century America, with their enormous wealth, were generous when it came to supporting science. In California, the astronomer George Ellery Hale succeeded in raising funds from the Carnegie Institute to establish a huge observatory on Mount Wilson, high in the San Gabriel Mountains above Los Angeles, and to install a 1.5m (5ft) telescope at its heart – the largest telescope in the world. Its "first light" was in 1908, but before it was even active, Hale

Right: The 100-inch Hooker Telescope was responsible for many important breakthroughs, including proof of galaxies beyond the Milky Way.

Left: Uranus (far left) and Neptune – a pair of worlds that are the farthest planets in the Solar System, discovered thanks to advances in telescope technology and mathematical astronomy respectively.

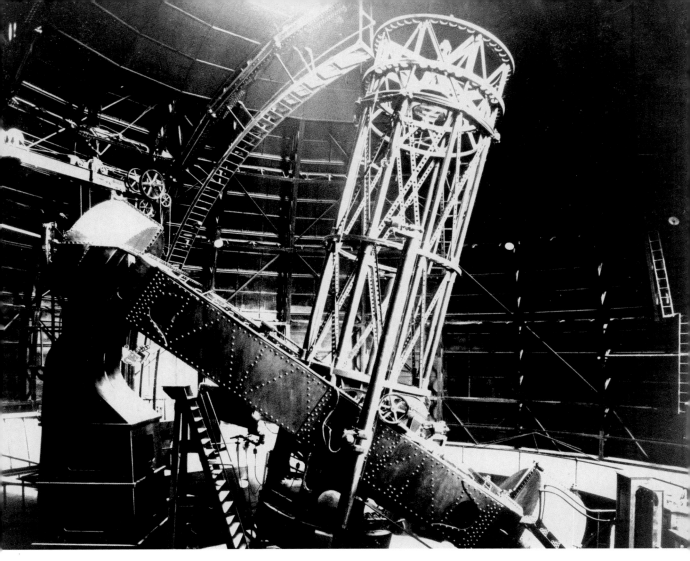

*"At the time
the Hooker
Telescope
came into
action it was
still a matter of
debate as to
whether or not
the Universe
extended
beyond the
Milky Way."*

was busy raising funds for an even bigger one. The mountings, housing, and 2.5m
(100in) mirror of the telescope were hauled piece by piece up the mountain road by
mules, and finally set to work late in 1917. It was known as the Hooker Telescope, after
the industrialist benefactor who paid for the giant mirror, and it too held the title for the
largest telescope in the world, which it retained until after World War II.

At the time the Hooker Telescope came into action it was still a matter of debate as to
whether or not the Universe extended beyond the Milky Way (what we now know as
our galaxy). Hale had hired a young astronomer called Edwin Hubble to work on the
"100-inch" telescope, and in 1923 Hubble observed a special kind of star, called a
Cepheid variable, whose luminosity varies in a precisely known way such that its
brightness or dimness can be used to provide a measure of how far away it is. The
Cepheid he observed was in the nebula of Andromeda, and Hubble calculated that it
was far too far away to be part of the Milky Way. Andromeda, and other nebulae he
observed, must be galaxies in their own right. Suddenly, the Universe was even more
vast than had been believed – we could actually see countless other galaxies around us.

BIG BANG

Hubble did not stop there. Astronomers by now no longer relied on direct visual observation – the Cepheids had been spotted on photographs taken through the telescope, which allowed for very dim objects to be seen over very long exposure times. By 1929, the 100-inch telescope was hooked up to a spectrograph, capturing light at different wavelengths, and it was this combination that allowed Hubble to make another extraordinary observation. He saw that the light from certain galaxies was very slightly redder than from others. It is a phenomenon known as "red shift", where light coming from an object that is moving away from the Earth will travel at a slightly longer wavelength. The longer wavelength shifts the colour more towards the red end of the spectrum (the spectrum that Isaac Newton had so convincingly demonstrated as forming the constituents of white light). It was clear evidence that parts of the Universe were rushing away from us. Hubble also observed that the further away the galaxies were, the faster they were moving away. We were in a rapidly expanding Universe. He was not the first to suggest that this might be the case, but here was tangible evidence that supported the theory of the "Big Bang" origin of the Universe, which had been proposed only a couple of years before. These historic observations by Hubble are what placed him in the astronomical hall of fame, and meant that his name was given to what is the most powerful telescope of all time, the Hubble Space Telescope, whose images from the furthest reaches of the visible Universe continue to amaze us all and show us, quite literally, what is out there.

Below: The Hubble Space Telescope was launched in 1990 to continue Hubble's investigations of the large-scale structure and history of the Universe.

EINSTEIN

In 1905, Albert Einstein published his theory of special relativity, which defined the speed of light as being constant and began the concept of space-time – it also gave us the famous equation $E=mc^2$, which shows that vast amounts of energy are contained within the tiniest mass (See Chapter Four), an understanding that is at the heart of atomic energy. In 1916, he published his general theory of relativity, which tied together space-time and gravity. He showed that space-time is warped by the presence of matter, and that the warping results in gravity, which makes matter move. In extending the theory to the Universe as a whole, he could only make it fit the prevailing view that the Universe was static by adding in a "cosmological constant" to his equations. Later, when it became clear that the Universe is in fact expanding, Einstein called his introduction of the constant his "biggest mistake" and threw it out. In subsequent years, many experimental measurements have backed up Einstein's theories, and they now lie at the heart of our description of how the cosmos works.

UNCERTAINTY

It is worth reflecting that the Hubble Space Telescope is placed in orbit only through our ability to get practical use out of Galileo's understanding of the mechanics of projectiles, Kepler's laws of planetary motion, and Newton's definition of gravity, but what it and other devices for measuring the make-up of the cosmos have uncovered is not quite as straightforward as those laws once suggested. Einstein's theory of general relativity, put forward in 1916, demonstrated that time and space and gravity can be, and are being, warped throughout the cosmos. And as if that wasn't bad enough, today's cosmologists readily accept the existence of black holes from which no light can emerge; they see the need for 11 dimensions of space-time; they speak of string theory, of membrane theory, of parallel universes, of dark matter, and dark energy, something that no one knows the nature of, but which must nonetheless exist. Yet for all of these seemingly weird phenomena, the mathematics is strong and convincing, however fanciful the ideas may sound.

"Our view of the night sky is no longer of a closed world, but of an infinite, and expanding cosmos of which we are only a very small part."

To the person in the street, modern cosmology offers answers to the question of what is out there that must seem as disturbing and unlikely as the idea that the Earth travels round the Sun must have seemed over five hundred years ago. What has changed is that today science tries to respond objectively to the evidence that it finds, even if that evidence goes against the received wisdom. Our view of the night sky is no longer of a closed world, but of an infinite, and expanding, cosmos of which we are only a very small part. But what is perhaps more important is that our whole outlook has changed. We have gone from seeing with a closed mind, to seeing with one that is constantly being forced to rethink – to make sense of what we find out there, in this terrifying but dazzling universe.

Left: In 1924, Edwin Hubble showed that the beautiful Andromeda Nebula was an independent star system more than 2 million light years away. Today we know that it is a spiral galaxy similar to our own Milky Way.

Right: The Hubble Space Telescope has delivered stunning images such as this portrait of the Crab Nebula.

Connections — Cosmos

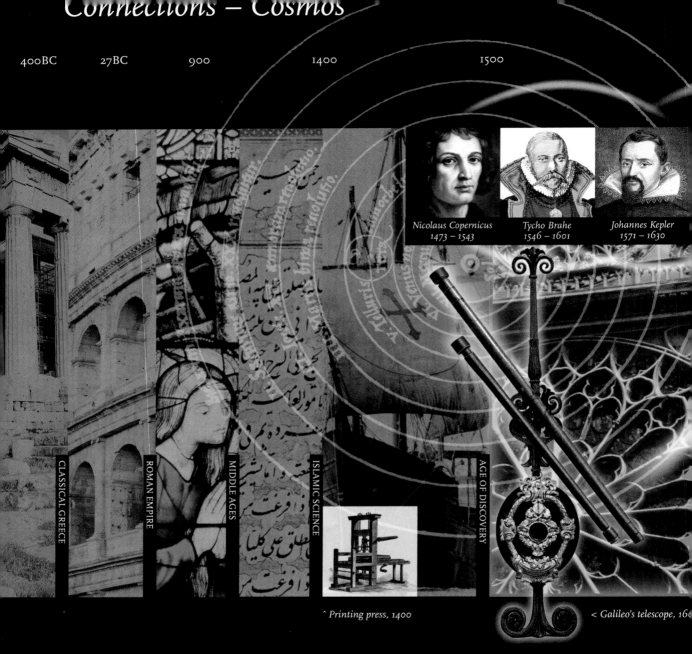

Nicolaus Copernicus
1473 – 1543

Tycho Brahe
1546 – 1601

Johannes Kepler
1571 – 1630

CLASSICAL GREECE

ROMAN EMPIRE

MIDDLE AGES

ISLAMIC SCIENCE

AGE OF DISCOVERY

^ Printing press, 1400

< Galileo's telescope, 16...

Our journey to discover what is out there has been shaped by powerful forces and beliefs. The ancient Greek view of a universe of perfect circles around the Earth was one of the most enduring ideas in human history. Dislodging it depended on accurate evidence gathered by inspired and ambitious men. Using a variety of instruments, including the most precise quadrants available, Tycho Brahe was able to compile incredibly detailed astronomical data over his lifetime. Johannes Kepler used this to prove mathematically Copernicus' earlier theory of a cosmos centred on the sun. But it was only when Galileo refined the newly invented telescope and began to observe the heavens that the new view became widely accepted.

The historical context was crucial too. The courts of the Renaissance encouraged new knowledge and paid for the study of the heavens. The religious upheavals of the Reformation, with two churches competing for control, created

Galileo Galilei
1564 – 1642

20TH CENTURY

Isaac Newton
1643 – 1727

Edwin Hubble
1889 – 1953

Albert Einstein
1879 – 1955

REFORMATION

AGE OF ENLIGHTENMENT

21ST CENTURY

^ *Tycho Brahe mural quadrant, 1600*

^ *Hubble Space Telescope, 1990*
^ *Mount Wilson Telescope, 1892*

an intellectual climate in which it became possible to question authority. The printing press allowed the new knowledge to spread at an unprecedented rate.

After Galileo, the search for ways to look ever deeper into the cosmos continued, as did the work of interpreting what was discovered, by brilliant men such as Isaac Newton. By the early 20th century, telescopes had reached a vast size. The biggest was the Hooker Telescope in the United States, where an astronomer named Edwin Hubble revealed that the universe was expanding. His name has been given to the most powerful telescope of all – the Hubble Space Telescope, whose images have revolutionized our view of what is out there.

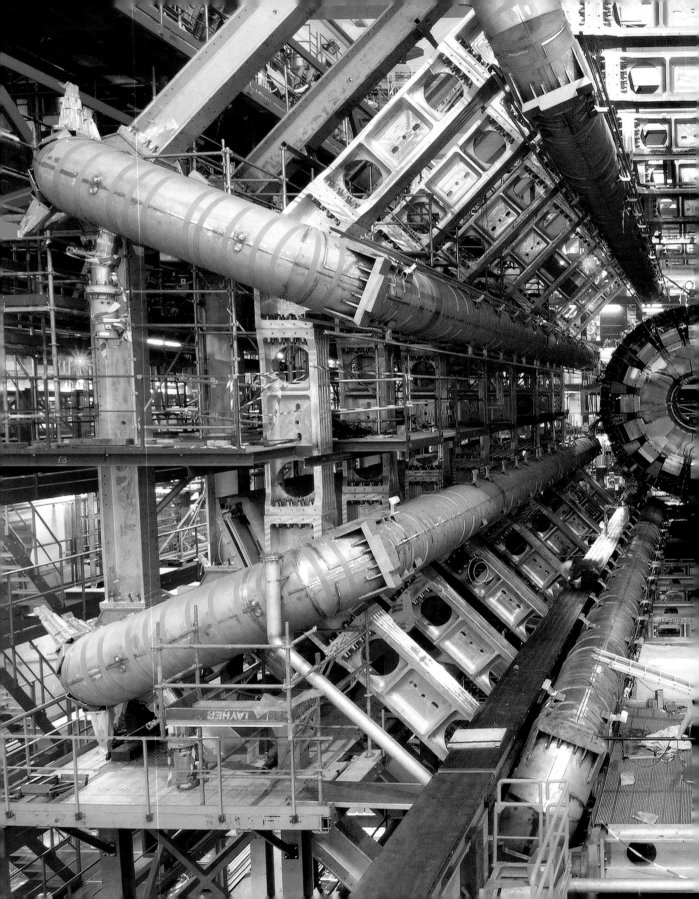

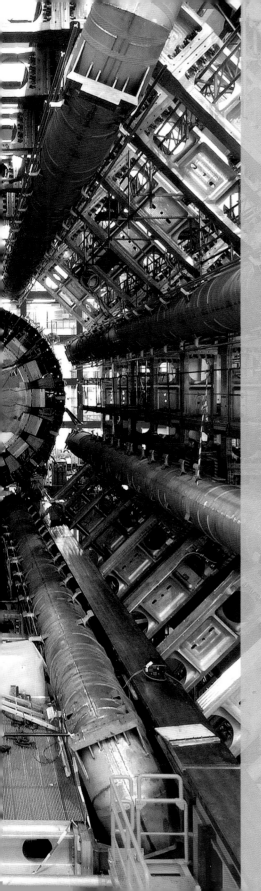

Matter

WHAT IS THE WORLD MADE OF?

The search to understand what the world is made of may seem, on the surface, to be a rather esoteric quest. Yet it has turned out to have had the greatest of practical consequences. Plastics, fertilizers, cars, computers, the internet, genetic engineering, the mobile phone, GPS – all have emerged, at least in part, from attempts to answer this question.

At the heart of it all lies our growing awareness of the way in which all substances, whether solid, liquid or gas, are ultimately composed of countless tiny particles – either molecules composed of even smaller units called atoms, joined together by chemical bonds to form simple or complex compounds, or lone atoms themselves.

Interactions between atoms and molecules forms the basis of all chemistry, but scientists had barely got to grips with this when they were confronted with evidence for another level of deep structure – a subatomic world that, paradoxically, both complicates and simplifies our models of matter. Instead of dozens of elements, most of the material world can now be understood in terms of just a handful of fundamental particles, but the way these few particles behave can be very strange indeed.

Nowhere is our quest to understand this subatomic domain better demonstrated than at the Large Hadron Collider, the world's largest science project, in Switzerland. Here, collisions between streams of particles travelling close to the speed of light aim to recreate conditions last seen in the Big Bang itself, releasing particles that may have stayed locked away since the creation of the Universe.

Left: A view inside the Large Hadron Collider, the world's largest and most ambitious particle accelerator, where streams of subatomic particles are collided with each other at speeds close to that of light in order to probe the deep structure of matter.

ATOMS

Today, it's common knowledge that the world, and everything in it, is made up of atoms – tiny, durable pieces of matter forged in the heart of stars. Atoms are very small indeed. A million of them stacked alongside each other would barely stretch across the letter "a" in this book. Each one consists of a nucleus containing protons and neutrons, around which swirls a cloud of electrons. The exchange and sharing of electrons between atoms in order to achieve stability is the basis of all chemistry, but reshaping atomic nuclei – either by breaking them apart (fission) or forcing them together (fusion) is the realm of atomic physics.

What's more, it's now clear that protons and neutrons are in turn made of even tinier particles called quarks, held together by the exchange of "messenger" particles called gluons. Who knows what further levels of structure still await discovery?

The secrets of the atomic and subatomic world remained hidden from us for centuries, by the limitations of our instruments and our thinking, and by the immense energies required to split successively smaller particles apart. So how did we learn what we now know, and how was the world transformed along the way?

"Atoms are very small indeed – a million of them stretched back to back would be about the size of the letter 'a' in this book."

INSIDE THE ATOM

The world is made of atoms of different sizes. They are measured in picometres, a unit equivalent to a trillionth (1/1,000,000,000,000) of a metre. A group of atoms can bind together to form a molecule, like water (H_2O), which consists of one atom of oxygen bound to two hydrogen atoms. Though they are often thought of as being rather like fuzzy tennis balls, atoms do not have an outside surface as such. Instead they consist of a tiny, dense, positively charged nucleus surrounded by a cloud of negatively charged electrons. Like the inside of a tennis ball, however, the atom is mainly empty space, which means that everything in the Universe, including us, is made mostly of nothing. Despite containing 99.9 per cent of the atom's mass, the nucleus, which is made up of protons and neutrons, is only a minute part of the whole atom – the scale is roughly that of a grain of sand to a cathedral. The number of protons in an atom determines its atomic number and therefore which element it is. A hydrogen atom, for example, contains one proton and so has an atomic number of one. Electrons, on the other hand, determine the chemical properties of an element. A helium atom has two electrons in its "shell", meaning the shell is full, and making it extremely stable and unreactive.

THE LAUGHING PHILOSOPHER

Democritus
c460–370 BC

Let us start with the Greeks and in particular the Laughing Philosopher, "the Mocker", Democritus. Democritus was born in Thrace in around 460 BC. As a young man, Democritus inherited a great deal of money from his father and spent it on travelling and broadening his experience of the world. He went to Asia, the Middle East, and possibly to Egypt. Somewhere along the way he picked up a love and knowledge of mathematics and also a thoroughly materialistic and deterministic view of the world. Democritus believed that everything could be explained by natural laws, if only he knew what these were.

Legend has it that Democritus was sitting at home when he smelt a loaf of fresh bread that a servant was carrying up the stairs. He started wondering how it was that he could smell the bread and concluded that minute particles of bread must escape into the air. He went on to describe a thought experiment with cheese. Imagine, he said, that you cut a bit of cheese in half, then in half again, and so on and so on. Finally you will arrive at something you cannot cut, which Democritus called "atoms". The table is hard, he said, and the cheese is soft because the atoms that make them up are either closely or loosely packed. He boldly stated: "Nothing exists except atoms and empty space; everything else is opinion." He rightly believed that atoms were small, numerous, and virtually indestructible.

He went on to claim that the Universe was originally made up of atoms that randomly flew around until they bumped into each other, coalesced, and formed planets, like the Earth. Furthermore, he said that the Earth is a giant sphere flying through empty space and that our Universe is just one of many. It is truly remarkable when you consider that he was making these claims nearly 2,400 years ago. Unfortunately, however, Democritus and his followers had no proof of the existence of atoms or any real support for these other theories. They were philosophical speculations, no more.

Above:
Democritus' cosmology of the Universe put the Earth and the planets at the centre, surrounded by stars and an "infinite chaos" of atoms.

Democritus lived till the age of 90, by which time he was blind. He is then said to have starved himself to death. But he lived long enough to realize that he had lost out in the battle of ideas to those of the younger and more persuasive Aristotle, who, building on the ideas of Plato, claimed that all matter was made of five elements: earth, air, fire, water and aether. These were controlled by "forces" of love and strife. Unlike Democritus's atomic theory, which suggested that all life, colour, and variation in the world could be explained by the interaction of tiny particles, Aristotle's theory gave things purpose. This idea that the world was not random but imbued with purpose was for some a more comforting way of understanding life. Perhaps it was because of this that the so-called atomic theory was abandoned for over two thousand years. Aristotle became such an influential thinker that his views on this and many other areas of science (a great deal of which were wrong) not only prevailed in Greece, but were later taken up by the Islamic world and became the cornerstones of alchemical and medical practice in Europe until well into the Middle Ages.

ALCHEMY

The question "what is the world made of?" now became an obsession of the alchemists. The earliest form of alchemy was developed in China well over two thousand years ago. It was based on an understanding of the world that was similar to Aristotle's. Differing proportions of fire, wood, metal and water explained why matter had different properties. Unlike the Greeks, who enjoyed philosophical speculation for its own sake, Chinese alchemists focused their efforts in very practical directions. Encouraged by a succession of emperors who feared growing old and dying, the Chinese alchemists set out to find the elixir of life – described by some as drinkable gold. Ironically, one of their early discoveries was gunpowder, which rapidly became an elixir of death. By the 12th century the recipe for gunpowder, along with many alchemical beliefs, spread to Europe where they would be developed and go on to shape the history of the world.

GUNPOWDER

Gunpowder is a mixture of sulphur, charcoal, and potassium nitrate (saltpetre), mixed roughly in the ratio 2:3:15. It is widely believed to have been invented by Chinese alchemists or Taoist monks sometime around the 9th century AD, though there are competing claims. The Chinese rapidly realized its military potential and used it to create bombs and fuel rockets that they used against the invading Mongols, though it was clearly not a decisive weapon as the Mongols successfully conquered China and founded the Yuan Dynasty at the start of the 13th century. The Arabs probably learnt the secret of gunpowder from Chinese traders and in the latter half of the 13th century Hasan al-Rammah put together a book of gunpowder recipes in his *Book of Military Horsemanship and Ingenious War Devices*. About the same time, the Europeans learnt of gunpowder, possibly from the Arabs. Roger Bacon, an English friar and philosopher, wrote: "By only using a very small quantity of this material much light can be created accompanied by a horrible fracas. It is possible with it to destroy a town or an army."

THE 17TH CENTURY

Let us now jump forward to 1695 and the laboratory – if it can be called that – of Hennig Brand, one of the last of the alchemists. Brand believed he was finally on the brink of discovering the legendary Philosopher's Stone, a sort of supernatural universal cleanser. Alchemists like Brand believed fervently in the Philosopher's Stone. It was said that the Stone could clean out the impurities from base metals and turn them into gold, the purest metal of them all, as it never corroded. Similarly, the Stone could clean the impurities from a person, ridding him of disease, making him immortal and pure of soul. It is understandable then, that so many men throughout the ages, including men of undoubted genius such as the physicist Isaac Newton, dedicated so much of their lives to finding it. Brand was no such genius but he was extremely determined. Having already exhausted the fortune of his first wife in his quest for the Stone, he began working his way through the even larger fortune of his second wife.

Brand decided he might find the Philosopher's Stone in human urine. It seems an unusual place to look and sadly we do not know what persuaded him to try. Alchemists like Brand believed that they were recovering lost secrets – fragments of knowledge about how to change the world, dating back to the ancients. They were obsessed with codes and ciphers, not only as a way of protecting their professional position, but also for keeping potentially dangerous secrets to a limited number of individuals. So although we know some of what Brand did, much remains a mystery. Perhaps he chose urine because it was gold coloured and readily available, or perhaps because it had already been found to have useful and unusual properties. The Romans used it to wash clothes – their launderettes, known as fulleries, stank of urine, which they had discovered was excellent for removing grease. Urine has also been shown to be a useful way of producing saltpetre, an essential ingredient of gunpowder. A 13th-century recipe claimed that the best source, or that which would produce the biggest blast, was Bishop's urine.

> *"Alchemists like Brand believed they were recovering lost secrets – fragments of knowledge about how to change the world, dating back to the ancients."*

What we do know is that Brand collected several barrels of urine from the local soldiers and left it to ferment; he estimated that he would need 5,000 litres (1,100 gallons) for his experiment. Using a complex series of glass flasks, pipes, and ovens capable of achieving high temperatures, he reduced and distilled it, then added sand – perhaps reasoning that it also had a golden colour – and distilled it again. Those who have attempted to repeat Brand's "experiment" have come to appreciate what a very able technician he must have been. To be able to boil, purify, and extract anything using such crude heat sources and under such dangerous conditions was

Above:
"Quicksilver", a metal that is liquid at room temperature, was associated with the Greek messenger god Mercury and was often represented by his symbol, the caduceus or snake-entwined rod.

Left above:
A page from a medieval work listing the supposed alchemical properties of various substances.

a truly remarkable achievement, let alone the smell. The white, waxy goo that remained after so much work must have been, initially, rather disappointing. There was no gold here, and certainly no Philosopher's Stone. But the goo did have a more remarkable property: it glowed in the dark. Brand named it "phosphorus", from the Greek meaning "light". Before long, Brand was touting phosphorus as an aphrodisiac and medical panacea, especially suitable for the treatment of mental conditions. Phosphorus appeared in 18th-century pharmacopoeias as a useful treatment, until it was recognized as being highly poisonous. It was then labelled "the devil's element".

We now know that Brand had isolated an element. He had made a significant discovery, but lacked any proper explanation of what was happening; he had the technology, and the methods, but not the science.

Initially, Brand had a great deal to be pleased about. Phosphorus was such a rarity that it was worth more than its weight in gold. An ounce of phosphorus sold for around six guineas. Not surprisingly, Brand tried to keep his production methods secret and for a while succeeded. But the volumes of urine required and the smell that the process produced meant it was inevitable that others would soon catch on.

"Phosphorus was such a rarity that it was worth more than its weight in gold."

MAKING PHOSPHORUS

It is ironic and tragic that phosphorus, which was first discovered in Hamburg, was used by the Allies in the bombs that later destroyed the city in World War II. The exact details of Hennig Brand's technique for producing it are not clear as he kept much of what he did secret, but broadly it consisted of leaving the urine to ferment and then boiling it down to a thick liquid or paste, which was then heated to a very high temperature, distilled, mixed, and heated again, until finally it yielded phosphorus. Brand needed around 1,000 litres (220 gallons) of urine to produce just 60g (2oz) of phosphorus. As it turns out, leaving urine to ferment was both odorous and pointless – just as much phosphorus can be obtained from fresh urine. His production method was also enormously inefficient, since one of the products he discarded after distillation contained much of the phosphorus present in urine. Brand's 1,000 litres (220 gallons) should have yielded over 20 times more phosphorus, since a single litre of human urine actually contains around 1.4g (0.5oz). Despite the cost and unpleasantness associated with its extraction, human urine remained the main source of phosphorus until the 18th century, when Carl Scheele, a Swedish chemist, showed that it could be extracted from bone ash.

Robert Boyle, a London doctor and the younger son of the Earl of Cork, was one of the first to uncover Brand's secret. He was told that phosphorus was derived from "somewhat that belonged to the body of man" and drew the correct conclusions. He made his phosphorus from urine, but unlike Brand realized its wider potential. Rather than poisoning his patients, Boyle used phosphorus to create matches. These were a great invention; the first reliable way to start a fire without need for flints to be cut, sticks to be rubbed, or eternal flames to be tended.

Boyle was, however, no mere match maker. Initially a believer in alchemy, he became deeply sceptical about its claims and was determined to make it more "scientific". In 1661 he published *The Skeptical Chmyist*, a book that finally reintroduced the idea of atoms, or "corpuscles" as he called them, to Western thought. Among many other things, Boyle experimented with compressing air by pouring mercury into a U-shaped tube of glass, sealed at one end. He discovered a relationship between the pressure he created and the volume of the trapped air. This "law", now known as Boyle's law, said that doubling the pressure would halve the volume of trapped air. Boyle concluded that the best explanation for this was that air consists of tiny corpuscles and that by adding pressure he was forcing them closer to each other. This suggestion was, however, not taken up and once again the idea of the atom faded into the background of mainstream thought.

We may laugh at Brand's belief that anyone could ever extract an elixir of life from human urine, but the fact is that alchemists like Brand paved the way for scientists like Boyle. Alchemists developed techniques, especially in reducing, separating, and distilling, that would prove essential for developing modern chemistry and a better understanding of the nature of matter.

Left: An illustration of one of Robert Boyle's most famous experiments, in which he used a vacuum pump to show that the sound of a ringing bell could only be heard if there was air to carry it.

THE CHEMICAL REVOLUTION

Moving on 80 years and by the middle of the 18th century gunpowder and cannons, discoveries of the Chinese alchemists, had been developed to truly terrifying levels of effectiveness by ingenious Europeans. Although the age of the great castle was over, brought to an end by the ability of cannons to shatter their defensive walls, religious differences fuelled almost continuous warfare across Europe and therefore the fate of nations still depended on the manufacture of cannons. The aim was to build bigger cannons, more cannons, and, crucially, cannons that would not explode in the operator's face. Artillery schools were set up in France and Britain to build a new class of technologically superior weapons.

"By the middle of the 18th century gunpowder and cannons had been developed to truly terrifying levels."

Making cannons required warring nations to develop their understanding of the metals needed to make them. This obsession with better metals sparked new interest in the processes of mining and turning ores into metal. Above all it brought about a new interest in gases, which was to be crucial in the next stage of attempts to understand matter.

HOW TO MAKE A CANNON

Although there is some evidence of types of cannon being used in the ancient world, the first that resemble what we would recognize today as a cannon almost certainly originated in China. They were simple tubes made of paper and bamboo, which were filled with gunpowder (another Chinese invention) and odd bits of shrapnel, and were almost as dangerous to the people who used them as they were to their enemies. Later, they were made of metal, iron, or brass. The Chinese mounted thousands of them on the Great Wall of China in a futile attempt to keep the Mongols at bay. From the Chinese, cannon technology seems to have been picked up by the Islamic world and then the Europeans. The arrival of the cannon in Europe transformed warfare, in particular the way that sieges were conducted. As Niccolò Machiavelli, a wily Italian political philosopher commented, "There is no wall, whatever its thickness, that artillery will not destroy in only a few days." By the time he was writing, in the early 1500s, people had realized that the longer the cannon barrel, the further it could fire. The result was that people started to build cannons that were truly enormous; cannons with barrels 3m (10ft) long and weighing over 9,000kg (9 tonnes) were not uncommon. As with the early bamboo cannon, however, there was considerable risk that the barrels would explode, killing the gunners.

THE DINNER

Antoine Lavoisier, the host of the dinner, was an urban sophisticate. He had inherited a small fortune, which he supplemented by becoming a so-called tax farmer. He paid the government a fee and in turn was given the right to collect taxes on the French government's behalf. Members of the "tax farm" were often corrupt and, perhaps understandably, extremely unpopular. Lavoisier would eventually be destroyed by association with this body.

At the tender age of 25, Lavoisier had been elected to the Academy of Sciences and was soon drawn to high-profile schemes such as designing better ways of lighting the streets of Paris. He had a young wife, Marie-Anne, who he had married when she was just 13, snatching her from under the nose of a much older man. She was his scientific soul mate and on most days they would spend five hours working together in his laboratory. On Sunday, which they called their day of happiness, they would try to spend all the day in the laboratory. Marie-Anne's role was vital in their joint work. She had learnt English so she could keep him up to date with work going on across the Channel. She created sketches of their experiments and kept detailed records of all their results.

Lavoisier was a hugely ambitious man, well aware of his many talents, but also well aware that despite all his money and all his efforts he had so far made no truly remarkable discoveries. It was Priestley's revelations at the dinner that took place in October 1774 that finally led him to carry out some of his most famous experiments. Priestley was perhaps a little overwhelmed by the occasion; by the fine food, the lavish setting, and by Antoine Lavoisier himself, who had money, ease, and such charm. Priestley told his wife in a letter that "most of the philosophical people of the city were present". During the course of the dinner, Priestley told Lavoisier all about his recent discovery, of the air with the fiery properties. He even told Lavoisier and his fellow dinner guests how he'd managed to create it. Across the table, Lavoisier listened intently; he knew this was something of great significance. Priestley noted how Mr and Mrs Lavoisier and the other dinner guests "expressed great surprise". As well they might. By talking so openly about his recent discoveries, Priestley was behaving in line with the best spirit of the age – of scientific openness – even though Britain and France were enemies. But he would, in time, come to bitterly regret having been quite so frank.

> *"During the course of the dinner, Priestley told Lavoisier all about his recent discovery, of the air with the fiery properties."*

Left: Pictured here with his wife Marie-Anne, Lavoisier was responsible for naming several important elements, including hydrogen and oxygen.

After the dinner, Lavoisier rushed off to repeat Preistley's experiment in his own well-equipped lab. This was when the trouble started. What happened next would turn the two into bitter rivals. Getting pure calyx of mercury was relatively simple – mercury salts were one of the few effective treatments for the pox, which was widespread in Paris. Lavoisier heated the calyx of mercury and collected the gas, just as Priestley had done. He discovered that the gas he collected had the properties that Priestley had

SPLITTING MATTER

The way that Henry Cavendish had first made hydrogen was by dripping acid onto metal and collecting the gas. The balloonists had simply scaled up the process. They filled a barrel with iron filings, added acid and water, and piped the hydrogen that formed into their balloon. There were two problems with this process. First, it created a lot of water vapour, which was a problem for balloons that were made of paper. And secondly, making the acid required saltpetre, and since saltpetre was also an essential ingredient in the production of gunpowder, there was not enough to go around.

Lavoisier's answer was elegant, simple, and cheap. Cavendish, discoverer of hydrogen, had shown that exploding oxygen and hydrogen together in an enclosed flask with an electric spark produced water. Lavoisier realized this meant that water is a compound made up of two elements – hydrogen and oxygen. He also realized that if he could reverse the reaction he could split water into hydrogen and oxygen.

In essence, what Lavoisier relied on was an accelerated form of rusting. He took a 1.2m (4ft) iron tube, a gun barrel, which he placed at an angle in glowing coals so that its central portion became red hot. He then trickled water down the barrel. As the water passed down the barrel, the oxygen in the water reacted with the metal of the gun barrel. This reaction left behind a mixture of steam and hydrogen gas. By condensing the mixture he was able to collect pure hydrogen. After perfecting his technique, Lavoisier repeated the experiment in front of an audience of over 30 gentlemen. He wanted them to see and take notice of what he had achieved.

The French government immediately recognized the military potential of hydrogen-powered balloons. They were soon being used on battlefields to keep track of enemy movements and Napoleon later set up the "Corps d'Aerostation" with the intention of using balloons to invade England. When that plan floundered, he took some with him on his invasion of Egypt. They never saw action as they were destroyed by Horatio Nelson, Vice Admiral of the English fleet, before they could fly. Hydrogen airships were later used in surveillance and bombing in World War One, but never quite took off as passenger aircraft, partly due to the iconic Hindenburg disaster, which destroyed public faith in their safety.

THE HINDENBURG DISASTER

The Hindenburg disaster is one of the most famous air disasters of all time. It took place on Thursday 6 May 1937 when the giant airship, while trying to dock in New Jersey, caught fire and was destroyed in less than a minute. Only 36 people were actually killed on the airship, making it in many ways a relatively trivial disaster; the speed at which the aircraft was destroyed, however, combined with the fact that it was being filmed as it burnt, made it an iconic moment in air travel. No one knows why the fire started, or indeed, where it started, but once the flames took hold, hydrogen lived up to its reputation as a highly inflammable, explosive gas. The irony is that the Zeppelins, of which the Hindenburg was one example, had a superb air safety record. No passengers, prior to the fire, had ever been injured, despite the fact that between them Zeppelins had flown many millions of miles. The destruction of the Hindenburg spelt the end of airships as a form of advanced passenger travel. Soon, aircraft replaced them as the most popular way of crossing the Atlantic.

Left: A popular school chemistry experiment demonstrates the explosive result when a balloon filled with hydrogen is ignited with a candle.

THE REVOLUTION

In 1789, the year that Parisians stormed the Bastille and launched what would come to be known as the French Revolution, Lavoisier published his masterpiece, *Elements of Chemistry*. The book was a summary of his life's work and swept away the last vestiges of the secret world of alchemy. Chemicals were given logical names – names that described what they did. The book is seen by many as containing the beginnings of modern chemistry. Unfortunately, Lavoisier did not live long enough to enjoy the fame that he clearly craved.

Lavoisier's lifestyle, and his chemistry, had been funded by his work as a tax collector. Tax collectors are not the most popular of people today, but in revolutionary France they were particularity hated and despised. Jean-Paul Marat, a leading member of the Revolution, who had been snubbed by Lavoisier, was one of those who now took his chance for revenge. "Lavoisier is a putative father of all discoveries that are noised abroad. Having no ideas of his own, he steals those of others; since he hardly ever knows how to evaluate them, he abandons them as lightly as he takes them up, and changes his systems like his shoes."

Accused of stealing money and counter-revolutionary activities, Lavoisier was guillotined in May 1794, aged 50. A contemporary commented sadly, "It only took a moment to sever his head, and probably one hundred years will not suffice to produce another like it." The night before he died Lavoisier wrote to a friend that he hoped he would be remembered "with perhaps a little glory". Maybe Lavoisier would be happy with how he is now regarded. He did not make any hugely original discoveries, but by interpreting the discoveries of others he contributed more than most to our understanding of the world.

Above: Lavoisier's *Elementary Treatise*, published in 1789 and illustrated by his wife, is widely considered the first modern chemistry textbook.

Left: Lavoisier was one of 32 former members of the "Ferme Generale" tax farm arrested in November 1793. Six months later, 28 of them were executed under the guillotine.

THE ROMANTIC CHEMIST

Humphry Davy
1778–1829

A few years after the death of Lavoisier, London acquired a new star. His name was Humphry Davy and from his first appearance on the stage in 1801 he drew brilliant reviews.

He drew such large numbers of the fashionable, young, and romantic to his talks that the street outside the Royal Institute, where he performed, was often a chaos of carriages. To cope with this the authorities made Albermarle Street London's first one-way street. Davy's huge success is surprising in light of the fact that he was a chemist and his performances were lectures about chemistry. But Davy certainly knew how to create a spectacle. As well as producing satisfying explosions he enthralled his audience with demonstrations of the new wonder of the age, electricity.

In 1799, Alessandro Volta, a professor of physics at Pavia University, had created the world's first battery (see Chapter Four). It was rapidly discovered that the battery, or voltaic pile as it was called, could be used to break down water into hydrogen and oxygen. Davy realized that this was an invention with which he could make his name and he raised money by public subscription to buy a large number of voltaic piles. One of his early experiments using voltaic piles involved applying intense electrical charge to an extremely caustic substance called potash. When he applied the electrodes to a container full of potash he created small globules of a substance that had "a high metallic lustre, being precisely similar in visible characters to quicksilver". This substance, which he called potassium, exploded and produced a bright flame when burned. The discovery of potassium thrilled Davy and delighted his audience when he repeated the experiment in public.

Using electricity to break down different compounds, Davy discovered a number of other elements, including sodium, calcium, strontium, barium, magnesium, and chlorine. He felt that he was uncovering a previously secret and hidden world. Davy realized that all matter is made of elements, an element being a substance that cannot be "decomposed by any chemical process". He also speculated that since electricity can prise matter apart, the force holding matter together might be electrical in nature. These were huge achievements.

"Davy certainly knew how to create a spectacle. As well as producing satisfying explosions he enthralled his audience with demonstrations of the new wonder of the age, electricity."

Left: The voltaic pile stacked several pairs of zinc and copper discs, separated by cloths soaked in brine. When the circuit was completed by a connecting wire, current flowed.

But perhaps his greatest achievement was in popularizing chemistry and firing the imagination of his audiences in a way that had never been done before. His demonstrations packed in young ladies keen to make his acquaintance. "His eyes", said one of his many female admirers, "were wasted on experiments".

But more importantly, his lectures also attracted budding industrialists and factory owners – serious men looking to learn something that would be useful to their business. Davy opened their eyes to the money that could be made from this new area of science.

Davy also took time to help a colleague, John Dalton, improve his presentation skills. He would make Dalton read out his speech while he sat at the back of the room and took notes. Dalton is now chiefly famous for yet again attempting to reintroduce the idea of atoms. Like Robert Boyle, he carried out a great deal of research on gases and became convinced that the pressure exerted by a gas on an enclosed container was due to atoms banging around inside it. He decided that every element was made up of its own, unique, kind of atom and that these atoms combined to form "compound atoms". These ideas were, however, largely ignored.

Davy died young, aged just 51. His health was probably undermined by the accumulated effect of testing on himself numerous toxic chemicals. Apart from his many discoveries (including a safety lamp for miners) he had another, rather subtler, effect on the development of science. Among the many women who followed Davy's lectures with such enthusiasm was Mary Shelley. When she wrote *Frankenstein* she based one of the characters in the book on Davy, and the idea of creating artificial life was certainly inspired by his demonstrations of the power of electricity.

Left: The violent purple flames created when potassium burns in air impressed both Davy and the audiences at his public lectures.

Shelley's book, and in particular the films that it later inspired, helped create in the public mind the image of a scientist as someone who is brilliant but also dabbling in forces they do not really understand and have little control over.

Another friend of Davy, the poet Coleridge, helped coin the word "scientist", to describe what experimentalists like Davy actually did. For many years "science" had been a blanket term covering a whole range of knowledge. You could talk of the science of music or the science of ethics, but the previous blanket term, "natural philosopher", was considered too lofty and "scientman" too ungainly.

THE ELEMENTS AND THE PERIODIC TABLE

The Greek philosopher Aristotle, who claimed that everything in the world was made of five elements, defined an element as, "one of those bodies into which other bodies can be decomposed but which is itself not capable of being divided into another". This idea, that an element is something that is indivisible, persisted until it became obvious that elements are in fact made up of much smaller atoms. The modern definition of an element is based around our understanding of atomic theory and it is now defined as a pure substance made up of a single type of atom. Among the more common elements are oxygen, carbon, hydrogen, nitrogen, iron, copper, gold, and silver. The number of known elements has, over the years, slowly risen. In the 18th century, Lavoisier created the first modern table of elements, listing just 33 in his book, *Elements of Chemistry*. Its brevity aside, this list would cause a few raised eyebrows today, because to those elements we still recognize, such as oxygen, he also added "light" and "caloric". We now know of 94 naturally occurring elements, as well as 23 manmade ones. The latest manmade element, named after Nicolaus Copernicus, is "copernicium". It has the atomic number 112 and the symbol "Cp".

A CENTURY OF SYNTHESIS

Up until the 19th century, chemists were at their most successful in breaking down matter into elements. Soon, however, they began to put elements together in unusual combinations, creating a world where artificial products began to dominate. The raw material out of which many of the new products were made was, strangely enough, coal. By the turn of the century, coal was not just being burnt to produce warmth and power, it was also converted into gas, which was then used to light homes. One of the by-products of gas production was coal-tar, an unpleasant viscous substance. In the 1820s, Charles Macintosh used it to waterproof cloth, creating the Mackintosh raincoat. His success encouraged others to look for further uses. August Wilhelm von Hofmann, first director of the Royal College of Chemistry in London, recognized that coal-tar was a complex combination of carbon, hydrogen, oxygen, and nitrogen, and was hopeful that it would be possible to rearrange these elements to synthesize quinine, the most sought after drug in the world. Quinine was the only known way of treating malaria and malaria was then, as now, one of the greatest killers on Earth. Von Hofmann did not succeed but out of his search would come – quite by accident – the mass production of chemicals, manipulating matter on an industrial scale.

Malaria, an ancient disease that had in the past led to the fall of empires, was by the 1800s a leading cause of death in the tropics and one of the greatest health problems faced by would be colonizers, like the British. Army officers sent back grim reports of the disease, which in India alone affected more than 25 million people and killed around two million people a year. No one was quite sure what caused malaria – the link hadn't yet been made to mosquitoes – but the superstitious believed that carrying a spider in a nutshell would cure it. The need to combat malaria became a national priority for Britain, and all the other countries with empires overseas.

The malaria drug quinine came from the bark of the cinchona tree, originally discovered on the slopes of the Andes in South America, where it was known as "the fever tree". Legend has it that an Indian, lost in the mountains and burning with fever, had found himself near a small lake ringed by the trees. He tasted the water, which was bitter and therefore considered poisonous, but by this point he was so delirious he didn't care. He drank, fell into a deep sleep, and when he woke the fever had gone. The secret of the tree was passed on by the Indians to Jesuit priests, who brought samples of bark back to Europe. Malaria was, at that time, endemic in Italy and a number of other Europeans countries and doctors soon found that "Jesuit powder" was particularly effective at treating it. The problem was getting enough bark to treat the millions who needed it; cinchona trees grow slowly and only in selective

Left: Before the late 1800s, malaria was often attributed to the poor air around marshes and swamps – it was only in the 1880s that it was linked to a parasitic infection carried by mosquitoes.

HOW MALARIA ENTERS THE BLOOD

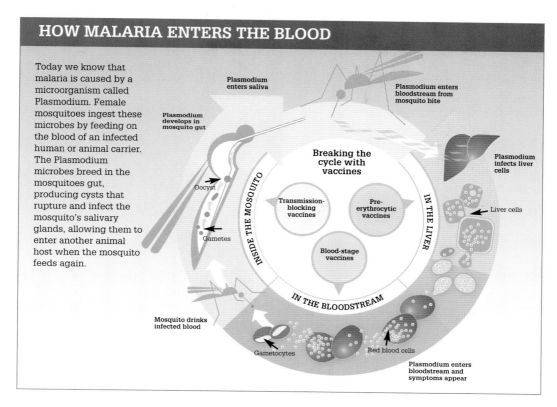

Today we know that malaria is caused by a microorganism called Plasmodium. Female mosquitoes ingest these microbes by feeding on the blood of an infected human or animal carrier. The Plasmodium microbes breed in the mosquitoes gut, producing cysts that rupture and infect the mosquito's salivary glands, allowing them to enter another animal host when the mosquito feeds again.

Plasmodium enters saliva

Plasmodium enters bloodstream from mosquito bite

Plasmodium develops in mosquito gut

Plasmodium infects liver cells

Breaking the cycle with vaccines

Oocyst

Liver cells

INSIDE THE MOSQUITO

IN THE LIVER

Transmission-blocking vaccines

Pre-erythrocytic vaccines

Gametes

Blood-stage vaccines

IN THE BLOODSTREAM

Mosquito drinks infected blood

Gametocytes

Red blood cells

Plasmodium enters bloodstream and symptoms appear

environments. To control malaria in India alone, the British would need 750 tonnes a year. Not surprisingly, there was a huge demand for a better and cheaper way to protect people against malaria. The British government turned to science. By the middle of the 19th century it was known that the active ingredient in the bark of the cinchona tree was quinine. The challenge was to find a way to synthesize it.

In the East End of London, a young man named William Perkin, one of the growing band of trained chemists, decided he was going to make his fortune by doing just that. Born in 1838, he attended the recently established Royal College of Chemistry and worked under the great von Hofmann, who encouraged him to see if he could make quinine from extracts of coal-tar. Perkin set to work in the attic of his parent's house. He started by trying to oxidize a substance derived from coal-tar called aniline sulphate. The result was an unimpressive black powder. But when he dissolved this in "spirits of wine" the results were spectacular. Perkin had discovered a completely new colour – mauve. He dyed a bit of silk and showed it to his friends, who were impressed and suggested there might well be money to be made from his discovery.

"The malaria drug quinine came from the bark of the cinchona tree, originally discovered on the slopes of the Andes."

The nearest alternative natural colour to Perkin's mauve was purple. It was a colour that had been used to dye the cloaks of Roman Emperors and was enormously expensive, since the only way to make it was from the glandular mucus of thousands of molluscs. Perkin's dye was far cheaper to make and in many ways superior – it always produced the same uniform shade, didn't smell of fish, and, most importantly, didn't fade in daylight. Perkin's mauve became all the rage. In 1858, Queen Victoria wore it to the wedding of her daughter, the Empress Eugenie, who was a fashion icon. Soon, the streets of London were awash with people wearing mauve – an outburst of "mauve measles". It brought colour to the Victorian age, and a knighthood to William Perkin. Mauve dye opened the floodgates to a surge of new colours, and a new industry was created to produce them. Dyes were one of the first chemicals to be made on an industrial scale and others, including fertilizers, soap, and dynamite, quickly followed. The manmade, the synthetic, took over from the natural.

> **"Perkin's mauve became all the rage. Soon, the streets of London were awash with people wearing mauve."**

The man who had inspired Perkin was the German, von Hofmann, and soon his former colleagues seized the chemical initiative. Trained German chemists came over to England, learnt Perkin's secrets and returned home. By 1878, the products of coal-tar production in England were valued at £450,000, while those of Germany topped £2 million. German chemists, working within a university system that actively encouraged research, had discovered a whole new range of synthetic colours. From being importers of natural dyes, Germany became the world's leading exporter of synthetic colour.

Another important German innovation was Fritz Haber and Carl Bosch's invention of a way to "fix" nitrogen from the air. This enabled large-scale production of ammonia, which led to artificial fertilizers. The resulting boom in food production led to massive population growth. Today much of the world would starve if it were not for artificial fertilizers. But artificial nitrates could also be used for explosives, and soon were. Meanwhile, the dye manufacturing process was producing large amounts of toxic chlorine gas as a by-product, which was of interest to the military for a different kind of weapon – poison gas.

Left: William Perkin made his great discovery of "mauveine" dye (below) aged just 18. He continued to develop new dyes and even synthetic perfumes throughout his life.

Right: A bottle of Perkin's original mauve dye, preserved in London's Science Museum.

ORIGINAL MAUVEINE PREPARED BY SIR WILLIAM PERKIN IN 1856

WORLD WARS

War was inevitable. Germany, France, and Britain were all competing for markets into which to sell their goods. They were expansionist and belligerent. Explosives were refined to a destructive power never seen before. Hydrogen balloons were used for observation and bombing. Then there were the gas attacks; the horrors of chlorine and other poisonous substances released on unprotected soldiers. World War I became known as the Chemists' War. The knowledge of how to manipulate matter had transformed the character of warfare itself. But even as the war was reaching its bloody climax, in laboratories in Germany and England the first elements for a new weapon of mass destruction were being put in place. The atomic bomb was the product of a 20th-century understanding of the world of matter, but its roots lay back in the 19th century.

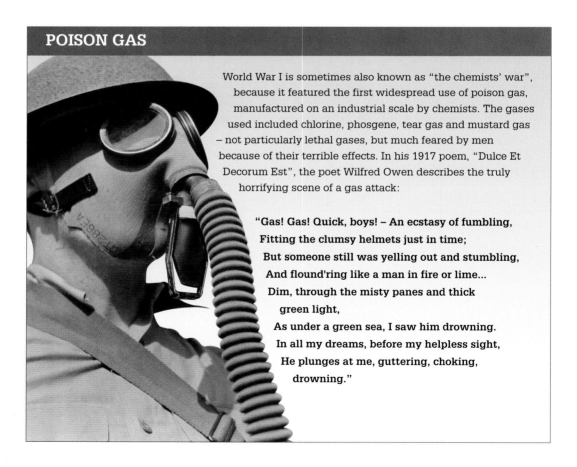

POISON GAS

World War I is sometimes also known as "the chemists' war", because it featured the first widespread use of poison gas, manufactured on an industrial scale by chemists. The gases used included chlorine, phosgene, tear gas and mustard gas – not particularly lethal gases, but much feared by men because of their terrible effects. In his 1917 poem, "Dulce Et Decorum Est", the poet Wilfred Owen describes the truly horrifying scene of a gas attack:

"Gas! Gas! Quick, boys! – An ecstasy of fumbling,
Fitting the clumsy helmets just in time;
But someone still was yelling out and stumbling,
And flound'ring like a man in fire or lime...
Dim, through the misty panes and thick
    green light,
As under a green sea, I saw him drowning.
In all my dreams, before my helpless sight,
He plunges at me, guttering, choking,
    drowning."

THINGS OF THOUGHT

"The man who in many ways initiated the next great revolution in physics and chemistry, a revolution that would usher in the atom and in doing so transform the 20th century, was William Crookes."

As we have seen, the idea that matter is made out of "atoms" – tiny indestructible balls – goes back many thousands of years. Every few centuries, the idea would reappear, only to be discarded again. No one, of course, had actually seen them, so their existence remained controversial. Ernst Mach, who gave his name to the units in which the speed of sound is measured, declared that they were no more than "things of thought". Mach, however, was in the minority. By the late 1890s chemists broadly agreed that hydrogen, helium, carbon, and all the rest were indivisible elements, each made up of its own unique atoms. It was hard to see where a new breakthrough would come from.

The physicists had also decided they had learnt most of what they were ever likely to know. Electricity, light, and magnetism could all be described by a few relatively simple equations. It was time to move on to more exciting areas of research. There were, of course, a few unresolved problems – like the Sun. According to existing theory, the Sun should have burnt all its fuel in its first 30,000 years. But the real death blow to scientific complacency came from a thoroughly unexpected source. The man who in many ways initiated the next great revolution in physics and chemistry, a revolution that would usher in the atom and in doing so transform the 20th century, was William Crookes.

Not exactly a household name, Crookes acquired some small fame for discovering the element thallium, but in his lifetime he was vilified by the scientific community for his other great passion – investigating spiritualism. Crookes, the eldest of 16 children, seems to have become interested in spiritualism soon after the death of his younger brother Philip, who died from Yellow Fever. He decided that as a scientist he had a duty to investigate all sorts of unlikely phenomena, arguing that this was the only way that anything new could ever be discovered. He went into his investigations a sceptic but was soon converted by what he saw, claiming to have witnessed levitation, phantom figures, and an accordion that played without human help. Crookes felt that he had found evidence pointing "to the agency of an outside intelligence".

Below: The spooky behaviour of the Crookes Tube, with its glowing walls and ability to spin a wheel with no apparent force, turned out to be early evidence for an invisible, subatomic world.

He realized that he could not explain how any of the things he witnessed actually happened, but pointed out that many scientists, himself included, were every day grappling with equally inexplicable phenomena in their laboratories. The scientific mystery that Crookes was puzzling over was a glowing glass tube. He and an assistant had built a vacuum tube, a sealed glass tube with an electrode at either end. Crookes discovered that when he attached a battery to the electrodes the glass wall of the tube became fluorescent. He decided that the rays that created the glow must be a stream of particles, as they were able to push round a tiny paddle-wheel that he carefully placed half-way up the tube. He was perplexed, however, by what he had found and called this phenomenon "radiant matter". He decided it was a new type of matter and was possibly connected in some way with the spirit world.

Other scientists, using improved and higher powered versions of the Crookes tube, soon began to make even more surprising discoveries. An undistinguished professor of physics called William Röntgen, who at the age of 50 must have thought his glory

days were behind him, noticed that a screen placed near his vacuum tube glowed every time he turned the power on. Following the maxim, "don't think, investigate", he discovered that the tube was producing invisible but high-powered rays that could pass through human tissue. Because he had no idea what these rays were he called them X-rays. The paper that announced this discovery to the world included an X-ray image of his wife's hand.

Left: An illustration by Crookes of the glow inside a vacuum tube, published in 1879.

Right: An early Röntgen X-ray photograph reveals the different X-ray absorbing properties of flesh, bone, gemstone and metal.

X-RAYS

X-rays are a form of electromagnetic radiation that readily penetrates human flesh. Almost as soon as the existence of X-rays was reported, the medical applications were recognized; as with radioactive materials, however, it took quite a while longer for people to fully appreciate the dangers of these mysterious new rays. Major John Hall-Edwards, who helped pioneer the use of X-rays in medical treatments, lost an arm due to excessive exposure, and there is no doubt that overexposure to X-rays has shortened many researchers' lives. As well as medical uses, X-rays have numerous other applications. In 1912, it was noticed that crystals cause X-rays to diffract, leading to the birth of a new area of research, X-ray crystallography, which in time would reveal the structure of DNA (see chapter Five) and rapidly advance the growth of modern genetics.

Not all X-rays are manmade, however. Extremely hot and compact stars, such as neutron stars and rotating black holes, also produce huge amounts of X-rays. Fortunately for life on our planet, these rays are readily blocked by the Earth's atmosphere.

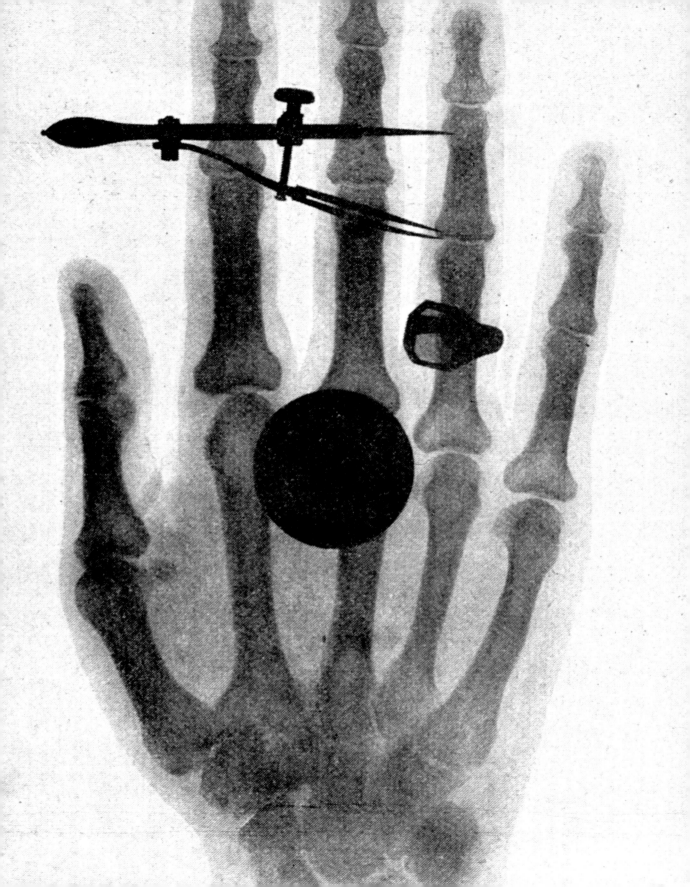

MYSTERIOUS RAYS

Henri Becquerel
1852–1908

Inspired by Röntgen's discovery, a French physicist called Henri Becquerel decided to investigate other phosphorescent materials to see if any of them also produced X-rays. His grandfather had been a famous collector of crystals that glowed in the dark, among them some uranium salts, so Henri used these for his experiments. He first "charged" the uranium crystals by exposing them to sunlight, then left them under some carefully wrapped photographic plates to see what would happen. Sure enough the plates subsequently showed the mark of the crystals. He repeated this experiment several times, then decided to put a copper cross between the crystals and the photographic plate. On previous occasions he had always charged the crystals with sunlight before doing the experiment. This time, either deliberately or because Paris was uncharacteristically gloomy, he did not. To his surprise the photographic plate showed the mark of the cross. The uranium salts were emitting mysterious rays, akin perhaps to X-rays, without needing to be charged.

Becquerel was clearly not that impressed with his findings, because he handed it over to one of his students, Marie Curie. Working with her husband Pierre, Curie soon discovered that pitchblende, the mineral ore from which uranium comes, produced huge amounts of energy without apparently losing mass. This was deeply puzzling because it seemed that these rocks were producing energy from nothing. In time it became clear that the pitchblende was unstable and was decaying to release different forms of radiation and huge amounts of energy. For her pioneering work Marie Curie won two Nobel prizes. She was also exposed to so much radiation during her research that it ultimately killed her. Her laboratory notebooks remain intensely radioactive, so much so that as historical artefacts today they have to be kept in lead-lined boxes.

Above: Working over four years, the Curies refined more than a tonne of pitchblende ore to extract just one tenth of a gram of a new element, radium.

Left: Marie and Pierre Curie, photographed at work in their laboratory at the School of Industrial Chemistry in Paris.

While Röntgen was investigating the mysterious X-rays produced by his Crookes tube, the physicist J.J. Thomson, working at the Cavendish laboratories in Cambridge (laboratories named after a relative of Henry Cavendish, the discoverer of hydrogen), had been studying Crooke's radiant matter. He agreed that the rays were made up of a stream of tiny particles, but his experiments suggested these particles were far smaller than the yet to be discovered atom. "The assumption of a state of matter more finely divided than an atom is a startling one," he declared. Thomson had found evidence of a subatomic particle – the electron.

THE ATOM

By the beginning of the 20th century it was becoming clear that the atom was not the smallest and most indestructible unit in the Universe, as people had been claiming. The Crookes tube and the discovery of radioactivity clearly pointed to the existence of particles that were far smaller than atoms. The problem was: how to study something so small that it could not even be seen. Working with two colleagues, Hans Geiger and Ernest Marsden, New Zealander Ernest Rutherford decided to use the atom to investigate the atom.

Rutherford had discovered that radioactive materials, as they decay, produce two very different types of radiation: alpha and beta particles. He used alpha particles, the heaviest and least penetrating, as his probes. Being notoriously clumsy, he delegated the actual experimental work to Geiger and Marsden, who spent days on end in a darkened room firing alpha particles at a thin sheet of gold foil. They could see where the alpha particles ended up by using a sheet of zinc sulphide, which sparkled when the alpha particles hit it. Most of the time the alpha particles passed straight through the gold sheet, but every so often a particle would be deflected. It was, said Rutherford, as surprising as if he had fired a 38cm (15in) shell at a sheet of paper and the shell had bounced straight back. Rutherford concluded from this that an atom consisted largely of empty space, with a small nucleus containing most of the mass. The positively charged nucleus, he said, was surrounded by tiny negatively charged particles – the electrons.

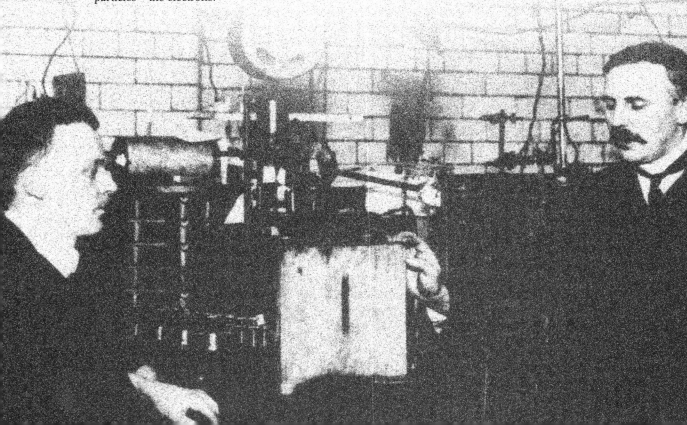

FLY IN A CATHEDRAL

To give some sense of scale, Rutherford imagined an atom blown up to the size of the Albert Hall. The nucleus, where almost all the mass of the atom resides, would be the size of a gnat – or as the newspapers put it, finding the nucleus was like trying to find a fly in a cathedral. This picture of a tiny fly buzzing around a vast cathedral is compelling but, in fact, quite wrong. Sticking with the example of a cathedral, the nucleus would actually be smaller than a grain of sand. Where Rutherford was right, however, was in claiming that the atom is mostly empty space. We, and everything in the world around us, consist almost entirely of void, of nothingness.

The mental picture that many of us have of an atom is that of a mini Solar System in which the electrons are like planets travelling through huge regions of empty space, revolving around the Sun, or the nucleus. This image makes sense and this was also how Rutherford first imagined the atom. But as other physicists soon pointed out, it was impossible that the atom could be anything like this. Unlike the planets, electrons carry a negative charge and classical physics predicts that a charged particle, when accelerating, will emit energy. If the electrons emitted energy, then they would soon collapse into the nucleus. But they clearly do not.

> *"We, and everything in the world around us, consist almost entirely of void, of nothingness."*

A colleague of Rutherford's called Niels Bohr became so obsessed with the problem that he spent his honeymoon working on it. His solution was so radical in its implications that many of his colleagues found it impossible to accept. Bohr himself once claimed that anyone who is not outraged when they first hear about quantum theory has not understood it. What Bohr said was that electrons only travel in fixed orbits; they are more like trams than buses. Every so often an electron will jump into an orbit that is closer to the nucleus and as it does so it emits a discrete package or quanta of energy. This is a so-called quantum leap. In fact, the word "leap" is misleading because what is being described is more like teleportation.

Left: Hans Geiger (far left) and Ernest Rutherford with their experiment in the laboratory at the University of Manchester.

Right: Simplified illustration of the Bohr model of the atom, with electrons moving on discrete and well-defined orbits around the nucleus.

Others rapidly built on Bohr's model, creating a description of the atom that is so far from common sense that it is hard for most of us to grasp. Within the world of the atom physicists had encountered an area of the Universe that our brains are just not built to understand. In the weird world of quantum mechanics, as it came to be known, electrons can be described

Electron in orbit   Nucleus   Restricted area

as both particles and waves. Their exact position is impossible to predict because until they are observed they do not really exist. They are probability waves. Electrons can not only teleport, they can, occasionally, pop in and out of existence. What's more, there are subatomic particles that demonstrate an ability that seems akin to telepathy. These particles have a twin with whom they are always in contact via a phenomenon known as quantum entanglement. If one twin is separated from the other and its spin altered, then the other will immediately start spinning in the opposite direction, no matter how far apart they are. Exactly how this happens is still unclear, but it invokes the possibility of instantaneous communication – information that can travel faster than light.

Quantum mechanics may seem abstruse, something best left for theoretical physicists to puzzle over; the subatomic world of probability waves, of electrons being everywhere and nowhere at once, seems unbelievably remote from our world. But the insights of quantum mechanics, among many things, provided the mathematical framework for the development of the transistor, a device that lies at the heart of the modern electronic world.

THE CONSUMER SOCIETY

The story of the splitting of the atom, leading to the atomic bomb and the development of nuclear power plants, is in many ways the story of a rather disappointing journey that has failed to live up to early hopes. It was once claimed that the atomic bomb would make the world a safer place and that nuclear fission would make energy so cheap that people would not even bother to meter it. Of course, none of these fanciful dreams has come true – at least, not yet. The story of the electron, on the other hand, has exceeded all expectations.

Imagine America in 1945. The war finally over, it was a time of optimism and pent up demand; the beginning of the long boom that would lead to the modern-day consumer society. It was also a world that was increasingly becoming dominated by vacuum tubes, also known as valves, remote descendants of the Crookes tube. The radio, for example – the main form of entertainment in the 1940s – was essentially a large clunky box full of valves. Telephones were growing in popularity, but telephone exchanges required lots and lots of valves. The first computer, Colossus, which cracked the Enigma Code and helped win the war, required more than a thousand valves. The valve was basically a device for controlling the movement of electrons. Valves were used to switch a current on or off, or else amplify a current. But the trouble with valves, besides being expensive and breakable, was that they used a lot of power and became extremely hot – to release electrons a lump of metal had to be heated inside a vacuum.

Enter William Shockley. At the end of the war he had influenced the decision to drop the atomic bomb on the Japanese by writing a report that claimed there would be a truly massive loss of lives if American troops tried invading Japan using conventional weapons. After the war he was employed by Bell Labs, the telephone company, to head a team trying to find a replacement for the valve. The business of

Above: Colossus, the programmable computer built at Bletchley Park in the 1940s, helped the Allies win World War II, but was then destroyed and kept secret for decades.

Following page: Shockley (centre) with Bardeen (left) and Battain. The three physisists shared the Nobel prize for physics in 1959 for their invention of the transistor.

connecting people was just taking off and Bell needed a better way of amplifying signals along telephone wires. The director of research at Bell was convinced the answer might lie in a strange new class of materials called semiconductors. He believed that electron behaviour in semiconductors could be manipulated to do the same job as the vacuum tube. Not only would this mean using less power, but it would also be faster.

"Without Shockley's transistor and the digital processors that it spawned there would be no portable computers, no mobile phones, no modern world."

Shockley was a man of undoubted brilliance, but also an extremely unpleasant boss. When, in late 1947, his team developed their first transistor, they kept Shockley's name off the patent. Shockley responded by locking himself away in a hotel room and within four weeks had designed a more rugged and practical transistor, the predecessor of today's transistors, which switch and amplify signals in almost every electrical device that exists today. Without quantum theory there would have been no transistor. Without Shockley's transistor and the digital processors that it spawned there would be no portable computers, no mobile phones, no modern world. In 1965, Gordon Moore claimed that the number of

SPLITTING THE ATOM

Splitting atoms is relatively simple; every time you switch on a light bulb you strip off electrons and split atoms. The real challenge is to split the nucleus of the atom. Ernest Rutherford, who, with his colleagues, revealed the structure of the atom by bombarding gold foil with alpha particles, tried repeatedly over many years before his first partial success in 1919. He had fired alpha particles (otherwise known as helium ions) at nitrogen gas, knocking protons out of the nucleus. The protons combined with fragments of nitrogen to produce small amounts of hydrogen. But the amount of disintegration was small and so this particular line of research reached a dead end. Fourteen years later, in April 1932, Rutherford's team at the Cavendish laboratories in Cambridge had far more success using a machine called a "linear accelerator", which produces accelerated protons. They fired the protons at a target made of lithium and found that the proton stream could split the lithium atoms into alpha particles. At the time, Rutherford and his colleagues had no idea that their research had any particular practical application, let alone that it would lead to a bomb more powerful than anything that had been seen before.

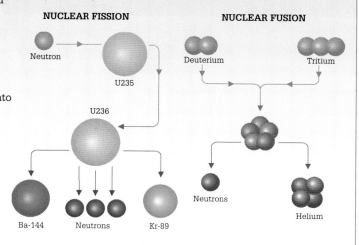

NUCLEAR FISSION

Neutron

U235

U236

Ba-144   Neutrons   Kr-89

NUCLEAR FUSION

Deuterium   Tritium

Neutrons

Helium

transistors on a chip would double every two years – Moore's Law. He was right. Intel, one of the world's leading transistor manufacturers, estimates that well over 10 quintillion (1 followed by 19 zeros) transistors are shipped every year. This equates to over a billion transistors for every human on earth.

William Shockley left Bell Labs in an atmosphere of bitterness and rancour after winning a Nobel Prize in 1956 and headed off to an orange grove near San Francisco. His reputation allowed him to recruit some of the finest minds in America, but soon they couldn't bear to work with him. Key men left the company but stayed on in San Francisco – and turned the orange grove into Silicon Valley, the high-tech heart of California.

SEMICONDUCTORS

Semiconductors lie at the heart of the transistor and therefore all modern electronic goods, such as computers, mobile phones, and televisions. The key to a semiconductor is the ability to control the rate of flow of electrons through it. Materials vary in how well they conduct an electrical current; metals, for example, tend to be good conductors, whereas rubber and plastics tend to be the opposite – good insulators. A semiconductor is a substance whose electrical resistivity – how strongly it resists an electrical current – lies somewhere between that of a conductor and that of an insulator. Critically, its resistivity can be modified by adding impurities (doping). Silicon, the most common element in the Earth's crust after oxygen, making up nearly a quarter, is widely used in the manufacture of semiconductors. This is why we talk about silicon chips and Silicon Valley, the place in San Francisco where transistor technology first took off.

Transistors are really just on–off switches. The semiconductor material in a transistor allows an electrical current to travel in one direction and control the size of that current. Running a small current through the transistor changes the electrical conductivity of the semiconductor, which then alters how much current passes through it. In this way, a small current can be used to control and switch much larger currents.

Above: A technician working on a detector at the European Organisation for Nuclear Reseach (CERN) laboratories, deep underground on the borders of France and Switzerland near Geneva.

THE FUTURE

Discovering the elements inspired a chemical revolution, which led to the development of medicines, plastics, and synthetic materials of every kind. Going down a layer took us to the atom. And probing the world of the atom has led, among other things, to nuclear energy, molecular biology, genetic engineering, molecular medicine, the computer, the laser, and the maser. It is estimated that the products of quantum theory account for about a third of the gross national product of industrialized countries.

Currently scientists at places like CERN, the European Organization for Nuclear Research and home of the Large Hadron Collider, are smashing particles together at extraordinary speeds in the hope of understanding the next level down – the quarks, which may or may not turn out to be the true building blocks of matter. A useful side product has been the World Wide Web, developed at CERN in 1989. The truth is we still don't know what the world is made of and it is possible we never will. The search, however, has been wonderfully productive.

400BC   27BC   900   1400   1500   1600   1700   1800

Democritus
c460 – 370BC

Robert Boyle
1627 – 1691

John Dalton
1766 – 1844

REFORMATION

CLASSICAL GREECE

ROMAN EMPIRE

MIDDLE AGES

ISLAMIC SCIENCE

AGE OF DISCOVERY

RENAISSANCE

^ *Crookes' tube*

As is often the story, the Ancient Greeks hit on both the right and wrong answers – and it was the latter that held sway. Democritus' theory that everything was made of tiny particles called "atoms" was eclipsed by Aristotle's idea of the five elements: earth, air, fire, water, and aether. This belief was echoed in the medieval practice of alchemy – the semi-mystical search to transform matter.

Driven by research into gases, Robert Boyle and later John Dalton attempted to reintroduce the idea of atoms, without success. Meanwhile, however, a revolution was occuring in chemistry – first in understanding and isolating elements, then in manipulating them to synthesize new materials.

iam Crookes
32 – 1919

JJ Thompson
1856 – 1940

Ernest Rutherford
1871 – 1937

Niels Bohr
1885 – 1962

AGE OF ENLIGHTENMENT

EARLY 20TH CENTURY

MID 20TH CENTURY

21ST CENTURY

ˆ First x-ray   ˆ Atom bomb   ˆ Atom model

By the end of the 19th century, most chemists agreed that atoms existed, but they thought that each element had
its own unique atoms and was indivisible. The breakthrough began with investigations into the mysterious glow
of William Crookes' electrified vacuum tube and the discovery of x-rays. J.J. Thompson realized that the rays were
made up of particles even smaller than atoms: electrons. Rutherford built on this insight to define the structure of
the atom, including the startling fact that it is mostly empty space. His colleague Niels Bohr solved the remaining
problem of how electrons orbit the nucleus, opening up the head-spinning world of quantum mechanics.

Understanding the structure of the atom paved the way for splitting it – nuclear fission. Rutherford's first partial
success with this happened in 1919; by 1932 he had succeeded and the countdown to the atomic age had begun.

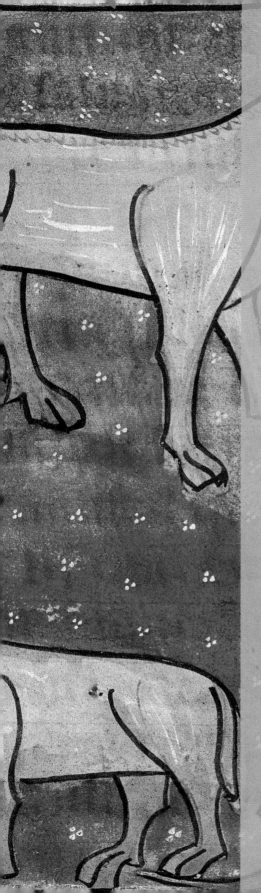

Life

HOW DID WE GET HERE?

In the late Middle Ages it was believed that the lion was the King of Beasts – its Latin name, *leo*, meant literally that. It was also believed that it used its tail to rub out its footprints and to deceive hunters, that it slept with its eyes open and that its cubs were born dead, only to have life breathed into them by their father after three days – just as Christ was said to have lain dead for three days before rising to heaven.

All this knowledge could be found in the "Bestiary", a list of the world's animals – initially all 40 of them – first compiled by an anonymous Greek after the 1st century AD, and added to occasionally over the centuries as it was studied by Arab and Christian scholars. Copied and re-copied, with beautiful illustrations that became ever more unrealistic with every reproduction, by the 1600s it was the definitive source of knowledge about the animal kingdom. As the number of creatures included grew they were grouped into "families", but these also included mythical beasts that no one had ever seen, including the phoenix, the dragon, and the unicorn. All creatures were imbued with religious moral significance, and their existence declared as being according to the divine plan. There must be a sea horse because there were horses on land, and divine symmetry required there to be one in the ocean.

And where did we fit in? The question of how we got here was one that no one in Christian Europe really asked. We sat at the pinnacle of life, at the end of a great chain of being that stretched down through animals, insects, and plants to the lowest forms of life, with everything fixed in its place, and we had been put in that position by the Almighty. But then people began to have the courage to go out, to look, to draw, and to study the natural world as it really was, and it was this that led inexorably to the realization that the question of "how did we get here?" was one that needed to be answered.

Left: A 14th-century manuscript illustration blends medieval interpretations of real animals such as elephants with fantastical figures such as wild men and dragons.

PRACTICAL PLANTS

The starting point of this quest was the discovery by Europeans of what they called the "New World" of the Americas. The aftershocks of this great event for human history are still being felt, but if we leave aside the quest for gold and silver, the invasions, conquests, wars, and battles for empires, the impact on the world of animals and plants was equally revolutionary. Over the next two centuries, thousands would make the long and hazardous journey from Europe to see the New World for themselves: adventurers, conquistadors, planters, and viceroys. But the discoveries also excited a sense of curiosity; a desire to see and experience the natural world in lands beyond Europe.

Hans Sloane
1660–1753

In 1687, an Irish doctor and botanist called Hans Sloane set eyes on Jamaica for the first time, after a three-month trip across the Atlantic. Sloane was there as the personal physician to the island's Governor, the Duke of Albemarle, but his primary interest was the natural world. He later wrote that he had come to the West Indies to "see what I can meet withal that is extraordinary in nature". In Jamaica he found more opportunity to indulge his passion for nature than he could ever have imagined.

But observing the natural world was much more than just a pleasant way to pass the day. In nature – and in particular plants – lay the foundations of imperial power. The discovery of the Americas had revealed a treasure trove of natural riches. Peppers, chillies, tomatoes, and the potato were just some of the botanical wonders to be brought back to Europe. Life on Jamaica itself centred around the sugar plant, the exploitation of which had turned the island from a pirate hangout into a cornerstone of the British Empire, with all of the terrible consequences for the slave populations that were carried there in their thousands to run the plantations. At their peak, Jamaica's estates were turning out 77,000 tonnes of sugar a year. And it wasn't just sugar; cotton and tobacco were shipped from the Americas, and tea and spices shipped from the East. Even the vessels themselves were built almost entirely from plants: oak for the frame and decking; cotton for the sails; hemp for the great coils of rope; and pine resin for the tar used to waterproof the hull.

Below: An early specimen drawer from Sloane's collection.

Global trade depended on plants and the search was on for more. Scientific curiosity ran hand in hand with potential profit, a motivation that was not lost on Sloane. The plant equivalent of the medieval bestiary was the "Herbal", similarly handed down from the work of a Greek living during the Roman era. The Herbal, however, also contained practical information – plants were listed according to their uses as foods or spices, or for their medicinal properties. The radish, for example, was described thus: "breeds wind and heats … it is good for such as desire to vomit." For millennia plants have been used as a means to treat illness and make our lives a little bit more comfortable, and the Herbal was

Above: A pressed cacao plant and seeds collected by Sloane in Jamaica, preserved alongside an illustration of the plant in one of Sloane's collecting albums.

effectively a catalogue of "simples", plants that each provided one medical drug. As a physician, Sloane had a keen interest in identifying new plants that might have medicinal value.

COLLECTORS

Sloane spent any time free from his duties as physician exploring the island on horseback, collecting and drawing its flora and fauna. He had been elected recently to the Royal Society and was one of a new breed of gentlemen of science who were obsessed by the idea of collecting and categorizing nature to fully describe and understand God's works. His work ethic was that of a strict Protestant upbringing. His employer, by contrast, indulged in drinking and carousing to excess and just 15 months after Sloane's arrival in Jamaica, the Governor died. Sloane had to embalm the body for the return voyage and then pack up his samples and go home. He had kept a meticulous journal of his time on the island, taking notes on everything he saw, from the weather to local customs and the geology of the island, and had gathered some eight hundred samples of flora and fauna, along with folios of detailed drawings of specimens that he had been unable to preserve. He even attempted to bring back with him a snake, an iguana, and a crocodile, but none of them survived the crossing. One plant specimen that Sloane picked up was the bean of the cacao tree, which he observed the local people turning into a drink for medicinal purposes. He described its effects as "nauseous and hard of digestion". However, he discovered that when mixed with milk it was quite pleasant and on his return to London patented it as drinking chocolate, promoting "its lightness on the stomach". Over the years it brought him a reasonable income, and it was Sloane's recipe that was eventually bought by the Cadbury family and turned into the successful industry it has become today.

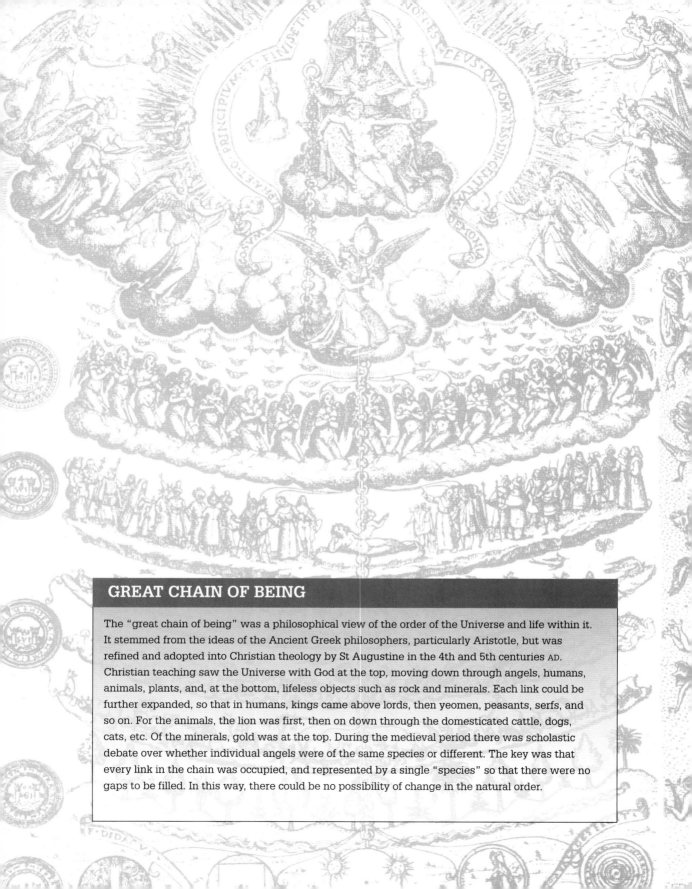

GREAT CHAIN OF BEING

The "great chain of being" was a philosophical view of the order of the Universe and life within it. It stemmed from the ideas of the Ancient Greek philosophers, particularly Aristotle, but was refined and adopted into Christian theology by St Augustine in the 4th and 5th centuries AD. Christian teaching saw the Universe with God at the top, moving down through angels, humans, animals, plants, and, at the bottom, lifeless objects such as rock and minerals. Each link could be further expanded, so that in humans, kings came above lords, then yeomen, peasants, serfs, and so on. For the animals, the lion was first, then on down through the domesticated cattle, dogs, cats, etc. Of the minerals, gold was at the top. During the medieval period there was scholastic debate over whether individual angels were of the same species or different. The key was that every link in the chain was occupied, and represented by a single "species" so that there were no gaps to be filled. In this way, there could be no possibility of change in the natural order.

CLASSIFICATION

Sloane's passion for collecting did not stop when he left Jamaica; in fact this was just the beginning. On his return to England, he began to acquire objects of an astonishing variety, even whole collections from others. By the end of his life, income from chocolate, his upper-class clients, shrewd investments, and marriage into a wealthy family had made him a rich man and helped him to amass the largest collection of curiosities in the world. There were 265 volumes of pressed plants; 12,500 more "vegetables and vegetable substances"; 6,000 shells; 9,000 invertebrates; 1,500 fishes; 1,200 birds, eggs and nests; 3,000 vertebrates, both skeletons and stuffed samples; and human "curiosities". There were minerals, rocks, and fossils in their thousands, jewellery, ornaments, medals, coins, art works, ethnographic objects, and a library of 50,000 books. The collection was so large and so special that, six months after he died in 1753, the government created the British Museum to house it. His collections formed the basis of the present day British Museum, Natural History Museum, and British Library, which have become three of the foremost storehouses of cultural, literary, and scientific knowledge in the world. And, true to his physician's calling, Sloane also provided the grounds for the famous Chelsea Physic Garden in London.

Yet Sloane's vast collection was just one example of what was happening across the globe. As more and more collectors brought back, or reported, more and more hitherto unknown forms of life it became clear that the description of God's creation that had been handed down for centuries was wholly inadequate. Sloane and his kind were overwhelmed by the huge diversity of what they saw. Even simple colours defied description, with Sloane commenting that they were so varied that there were insufficient names to note them all down.

John Ray
1627–1705

What was needed more than anything was the means to precisely distinguish one kind of living thing from another. The system of classification that Sloane had endeavoured to use in Jamaica and in his subsequent collecting was that espoused by the naturalist John Ray. The son of a blacksmith, Ray was influenced by his mother, a herbalist who used plants to treat the sick. He was a very bright boy at school and studied at Cambridge University, before being ordained as a priest. He was then caught up in the political and religious turmoil following the English Civil War and restoration of the monarchy in Britain, and found himself, as a priest, unable to take up secular work, but also unable to practise because of his nonconformist principles. Fortunately he was taken under the wing of an aristocratic friend from Cambridge, Francis Willoughby, and together they toured Europe for three years, studying flora and fauna as they went. Ray joined Willoughby's household and, after Willoughby died, devoted some years to completing a book about fish that his friend had started – the same book that almost bankrupted the Royal Society at the time Isaac Newton was trying to persuade them to publish *Principia* (see Chapter One) – before publishing his own substantial work on botany.

Ray's classification of some 18,000 plant species was based on their habitats, distribution, morphology, and physiology. It was the first seriously scientific attempt to bring order to the natural world. But his crucial contribution was to define the concept of "species" in its modern sense, with the notion that "one species does not grow from the seed of another species". It was this idea that was taken up by an obsessive and egocentric Swede called Carl Linnaeus and turned into the classification system for plants that is still in use today.

Linnaeus is always known by his Latinized name, although he reversed his father's adoption of that form when he was granted the status of nobility, and called himself Carl von Linné. Born in 1707, he was always interested in plants and was nicknamed "the little botanist" as a boy; he studied medicine and then became a lecturer in botany at the University of Uppsala. In 1732, he was financed to undertake an expedition through Lappland, from which he reported dramatic adventures and extraordinary hardships, although in reality his journey was significantly less strenuous or extensive, and he "borrowed" descriptions of places and even his itinerary from the diaries of a previous explorer. Linnaeus had an obsessive desire to order and systematize everything, but it was a personality trait that was perfect for setting the science of botany on a firm footing. He latched on to the suggestion that plants reproduced sexually – something that had not been considered before – and created a system of classifying them according to their complement of male and female sexual organs, known as stamens and pistils respectively. This shocked 18th-century polite society;

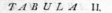

Above: Linnaeus' work on the classification of nature grew from a mere eleven-page pamphlet to a 3,000-page, multi-volume encyclopaedia by the time of its thirteenth edition.

SYSTEMA NATURA

The system that Carl Linnaeus devised for classifying life in *Systema Natura* was based on a hierarchical structure: three kingdoms, which were subdivided into classes, then into orders, then genera, and finally species. Everything had a place.

The big advantage was that it meant that the naming of an animal or plant also provided a definition of where it stood in the order of nature. Man was in the kingdom Animalia, of the class Mammalia, the order Primate, genus Homo, and species Sapiens – hence *Homo sapiens*. Linnaeus introduced the name of the class Mammalia, after the mammary glands, because all members of that grouping are animals that nurse their young. By the time of the tenth edition of Linnaeus's book, it classified over 12,000 species of animals and plants. It is a reflection on the beliefs of the time, when mythical animals were thought to really exist, that within the genus Homo he also included other species of man such as troglodytes (cave dwellers), mountain dwarves, and other such creatures. Today, Linnaeus's system has been refined to include more categories of branching in the tree of life, but the neatness of the two-word naming system has stuck, and remains an essential tool of biology.

his descriptions of plant reproduction were somewhat racy, using phrases such as "twenty males or more in the same marriage". He also invented the two-word (binomial) Latin naming system that is still used today. His *Systema Natura* was published when he was still not 30 years old and he continued to refine it and apply it to the animal and invertebrate kingdoms in editions published over the next 20 years.

For a man who made lists of everything in life – Linnaeus even classified other botanists – it was inevitable that human beings were also included in his plan. Linnaeus was the first to place Man amongst the beasts, and was acutely aware of the potential criticism he would be open to. For all that, Linnaeus believed firmly, as did almost everyone at that time, in the immutability of species. But the vast numbers of newly discovered and categorized life forms raised uncomfortable questions about God's plan. It seemed bewildering to be confronted with so many apparently pointless variations on a theme – why would God have created quite so many different species of beetle, or fish, or flowering shrub? But there they were. They did not change. They had always been there. They did not come and go. All of that certainty, however, was about to be undermined, quite literally.

Above: In Linnaeus' time, fossils such as this ammonite were explained as unusual mineral formations or remnants of animals from before the biblical Flood.

HOW FOSSILS FORM

The name fossil comes from the Latin meaning "dug from the ground". Until the 17th century, most people thought fossils were simply unusual stones – they did not realize that they could provide valuable information about the distant past. We now know that fossils are the remains of living organisms that have been preserved in sedimentary rock after they have died. They form when a dead creature or plant is covered by layers of silt or sediment. As the sediments solidify to hard rock, very slowly, molecule by molecule – and all the while retaining its structure – the organic material is replaced by minerals from the surrounding sediment. If a creature is covered quickly, before scavengers can damage the corpse, the fossil will be more completely preserved. Former river beds and volcanic ash fields are therefore good sources of well-preserved fossils. Sometimes the living tissue decays completely and leaves only an imprint of the shape in the surrounding rock. These are called trace fossils.

STRATA SMITH

Above: Francis Egerton, the 3rd Duke of Bridgewater, became obsessed with the idea for a canal to transport coal from Worsley to Manchester. He followed it up with an even more ambitious scheme linking Manchester and Liverpool.

By the second half of the 18th century the Industrial Revolution had begun to take hold in Britain. The textile and manufacturing industries were becoming mechanized, with machines like the spinning-jenny and the spinning mule; mass production had been brought into the potteries with Josiah Wedgwood's new factory, called Etruria, in Staffordshire with mechanized potters' wheels and lathes. New techniques in iron and steel production saw dramatic new structures made of metal being built. And just beginning to emerge as the great power behind the industrial growth of Britain was the newly developed steam engine. But more than anything the Industrial Revolution was founded on the mining of coal, and the transporting of that coal to the industrial centres that needed it: coal for blast furnaces and iron puddling; coal to power the steam engines; coal to heat the chimneys in the growing numbers of homes in industrial towns. It was the shifting of coal throughout Britain, along with the heavy goods produced by industry, which spurred a massive programme of canal-building across the country. The opening of the Duke of Bridgewater's canal in 1761, running from his mines at Worsley to Manchester some 10km (6 miles) to the east, halved the price of coal in the city. There followed a period known as "canal mania", and over the next 60 years more than a hundred canals were built across the British Isles, carrying coal to fuel industry, carrying the products of factories to the ports for shipment overseas, and carrying the raw materials that were flooding back from the plantations abroad. All part of the "virtuous" circle of trade that fuelled the growing British economy.

William Smith was born of an Oxfordshire farming family, and in 1787 at the age of 18 began work as an assistant surveyor. He eventually found himself working in the coal fields of Somerset, and on the construction of the Somerset Coal Canal. In the pits he began to recognize that the rocks appeared to be in distinct layers, or strata. The principle that sediments are laid down horizontally, but then somehow deform to run at different angles had been established a century before, but what Smith realized was that the layers seemed to run continuously, and that different strata could be traced across a distance. The key to his insight were imperfections that he recognized in the rocks – fossils.

Left: The 16km (10 mile) long Bridgewater Canal was actually built by James Brindley, one of the greatest engineers of the time.

Right: William Smith's hand-coloured geological map of Britain established the principle of using different tints to represent different rock units, and still bears a striking resemblance to modern maps.

As a boy Smith had always been interested in fossils, and recalled playing with shiny brachiopod fossils as marbles. Now, in the rock strata of the mines and cuttings carved out by the canal workers, he saw that each layer of rock could be recognized by the nature of the different fossils that it contained. So although the rock strata might now run at an incline, or even fold, dip out of sight or crumple or shear, their characteristic fossil markers enabled them to be traced across the land, wherever they appeared. On the basis that younger rocks must surely lie above older ones, it meant that the characteristic fossils could be used to provide the relative age of the rock layers – although of course at that time there was no way to obtain an absolute date for the age of the rock. Smith called it the "principle of faunal succession". He travelled thousands of miles across the British countryside in his role as a surveyor in the last decade of the 18th century, and his notes and observations enabled him to publish a stratigraphic map of Britain, the first of its kind, in 1815, a vast work, with each layer of rock beautifully coloured, offering a clear vision of the potential mineral worth of different parts of the country. Smith himself had a tumultuous life. His stratigraphic work was plagiarized, he lost money, and spent years in a debtors' jail. But before he died his contribution to science was recognized, and he is remembered as the "father of English geology". His glorious map today hangs on the wall of the Geological Society in Piccadilly in London.

"In the rock strata of the mines and cuttings carved out by the canal workers, Smith saw that each layer of rock could be recognised by the nature of the different fossils that it contained."

A
DELINEATION
OF THE
STRATA
OF
ENGLAND AND WALES,
WITH PART OF
SCOTLAND;
EXHIBITING
THE COLLIERIES AND MINES,
THE MARSHES AND FEN LANDS ORIGINALLY OVERFLOWED BY THE SEA,
AND THE
VARIETIES OF SOIL
ACCORDING TO THE VARIATIONS IN THE SUBSTRATA,
ILLUSTRATED by the MOST DESCRIPTIVE NAMES
BY W. SMITH.

THE

GERMAN

OCEAN

FIRTH OF FORTH

SCOTLAND

THE

IRISH SEA

ST. GEORGE'S CHANNEL

CAERNARVON BAY

CARDIGAN BAY

BRISTOL CHANNEL

EXPLANATION.

ENGLISH CHANNEL

HOT ROCKS

The first objective, scientific attempt to date the planet can probably be said to be the work of a geologist and naturalist from France with aristocratic pretensions, Georges-Louis Leclerc, Comte de Buffon. Born plain old Georges-Louis Leclerc, he inherited a vast fortune from his mother – including the village of Buffon, near Dijon – and thereafter, from the age of 25, called himself de Buffon. His geological experiment was based on speculation by Isaac Newton in his *Principia* (see Chapter One) about the rate of cooling of comets, which Newton had observed sometimes fell into the Sun. Buffon had the notion that such impacts might throw bits of hot Sun out into space and thus that the Earth started life as molten material spinning around its star, gradually cooling to the point at which it could sustain life. He reckoned that if he could work out how long it had taken for the Earth to reach this cooled state, he could estimate the date of its creation.

Buffon had his own forge make a series of iron balls to represent the Earth, ranging from about 1cm (0.4in) in diameter up to about 15cm (6in). He put them into the hearth, one after the other, until they were red hot. Then he took them out and timed how long it took before they were cool to the touch. In a description of his experiment it is recorded that he "had resort to four or five pretty women, with very soft skins ... and they held [the balls] in turns in their delicate hands, while describing to him the degrees of heating and cooling." By timing a range of different balls he was able to extrapolate his results to a globe the size of the Earth. When Buffon performed this experiment using several different-sized balls he concluded that it had taken a total of 42,964 years before the planet could sustain life, and an overall figure of about 75,000 years to reach its present day temperature.

Make no mistake, this was radical. Such a figure was an unimaginably long time by the standards of the day, and it was a result that drew ire from theologians at the University of Paris. But when Buffon published his figure, in 1749, he was careful to avoid any direct challenge to scriptures, and his *A Theory of the Earth* formed the first part of his monumental life's work, a description of the whole of the natural world, *L'Histoire Naturelle*, in 44 volumes. This series of books laid out a vision of the Earth that had a totally new perspective. Buffon argued that both the Earth and the life on it had a history and that he had identified seven great epochs of that history, which conveniently formed a metaphor for the seven days of creation. Above all, though, he suggested that species could change, with time and by migration from one part of the globe to another. Yet Buffon's experiments on time were just the beginning.

Right: Appointed head of the Jardin du Roi (royal gardens) in Paris in 1739, Buffon transformed them into a centre for research, paving the way for successors such as Cuvier.

Above: A pair of bird illustrations from an 1829 edition of Buffon's *Complete Works*.

LETTRES
PATTENT
DU ROY

HISTOIRE
...RELLE

DEEP TIME

If you travel along the south-eastern coast of Scotland from Dunbar towards Berwick-upon-Tweed, and take a boat out from the small coastal town of Cove, you will be greeted, a couple of kilometers further on, by a remarkable sight: the rocks of Siccar Point. In 1788, just such a boat trip by one James Hutton provided confirmation of an idea that had been forming gradually in his mind, and which was both startlingly clear but also shocking in its implications. Siccar Point shows horizontal strata of red sandstone overlying vertical layers of much older rock, known as greywacke. His interpretation was that the vertical stripes had themselves been laid down horizontally, just as all sediments are, in a deep sea environment, but gradually had been tilted through 90 degrees by huge forces within the Earth. The newer horizontal layers of sandstone had been laid down on top as part of a coastal plain that had suffered repeated flooding and then drying as sea levels rose and fell. To Hutton this was evidence that the processes of geology were slow, continuous, and had always been going on in the same way. The location is still known as Hutton's Unconformity.

The son of a merchant, Hutton had been apprenticed in law, dropped out, and then tried medicine before following a commercially successful career by pursuing his interest in chemistry. He became wealthy with the invention of an industrial process to make ammonium chloride from soot. He had also inherited a farm, and the outdoor life it brought encouraged his interest in geology. His industrial wealth enabled him to devote his later years to this passion, and he made many journeys around England and Scotland observing the landscape and building up his theory of gradual change. He saw signs of slow, steady erosion, successive earthquakes and uplifting of rock strata, and evidence of repeated volcanic eruptions in the past. Everywhere he looked, Hutton saw evidence that the natural forces at work on the planet today happened very gradually and that those same forces had always been at work for all time.

"Everywhere he looked, Hutton saw evidence that the natural forces at work on the planet today happened very gradually and that those same forces had always been at work for all time."

Below: An illustration of the geological "unconformity" discovered by Hutton at Siccar Point in Scotland reveals the abrupt transition between the shallow-sloping overlying rocks and the vertical strata beneath.

This theory of gradual change, which did not need sudden dramatic catastrophes – Biblical or otherwise – was known as Uniformitarianism. What it did imply, however, was an Earth that was far older than any estimate hitherto. As Hutton famously wrote in his book, *Theory of the Earth*, "we see no vestige of a beginning – no prospect of an end". The stage was set for an intellectual battle.

Charles Lyell
1797–1875

Hutton's published written style was almost incomprehensible, but his work was brought to wider attention a few years after his death in a book written by his friend John Playfair, who had shared the boat trip to Siccar Point. Uniformitarian thinking was roundly condemned by both Catastrophists and Neptunists before another Scot, who also abandoned law as a career, definitively demonstrated the reality of gradual change in the landscape.

Like Hutton, Charles Lyell followed his parents' wishes in studying to be a lawyer, but was inspired at Oxford University to follow geology by the Reverend William Buckland, an eccentric don who, among other things, had set himself the lifetime goal of eating each of God's creations and had a reputation for startling his students by offering bizarre new "delicacies" for tea. Buckland was very much a Catastrophist in the Cuvier mould and felt he had found evidence of the flood itself when, on a trip to Yorkshire, he discovered fossilized Hyena bones in what he interpreted as a muddy den. But around the same time Lyell was also introduced to Hutton's idea of uniformity and saw that it chimed with his own thinking; a family tour of Europe as a 20-year-old student, in the summer of 1818, further whetted his appetite for studying the landscape. He went on to qualify and even practise briefly as a barrister, before trouble with his eyesight forced him, probably not without relief, to abandon a career that involved intensive study of paperwork. Instead, he began to write essays and reviews for a living while pursuing his geological interests. The more Lyell studied the landscape, the more he became convinced by Hutton's concept of very gradual change, and an extended field trip to Italy in 1828 provided two startling pieces of evidence in its favour.

At Pozzuoli, near Naples, there are impressive Roman ruins are even more impressive for what they reveal about the changing Earth. At the Temple of Serapis, Lyell saw three tall columns still standing, each marked with a dark band of discolouration near the top. On closer inspection, the dark bands were marks of erosion caused by marine molluscs. The only possible conclusion was that after the temple had been built, the sea had risen – or the land subsided – and stayed there long enough for the small aquatic creatures to leave their mark, before the land rose again to lift them high above the water line. What's more, all of this must have happened very gradually, or the columns themselves would have been toppled. This was stark evidence of gradual geological change. But what Lyell saw in Sicily at the spectacular volcano, Mount Etna, was even more shocking. Looking out across Etna, he saw that it was built up from successive flows of lava and that an inordinate amount of time must have passed for the great height of the mountain to have built up layer by layer. Lyell realized that the Earth must be immeasurably old, and that the processes that we see in action today must have been at work for all time.

His book, *Principles of Geology*, was published in three volumes in 1830–33, with the Temple of Serapis as its frontispiece. In it, Lyell used his gift of a barrister's power of argument to counter and demolish the catastrophists' interpretation of the evidence in the rocks. The concept of deep time became hugely influential, and the poet Alfred Lord Tennyson was inspired by it when he began writing *In Memoriam*, after the death of a close friend:

"There rolls the deep where grew the tree.
O earth, what changes hast thou seen!
There where the long street roars, hath been
The stillness of the central sea."

Tennyson's words evoke a timeless planet, but also one that continues with no apparent purpose or direction. Lyell's arguments had set the stage for the realization that such vast amounts of geological time could allow for gradual change in life itself. The concept of evolution was in the air.

Following page: In the 1830s, the Temple of Serapis attracted the attention of another famous scientist, mathematician Charles Babbage, who spent some time there and compiled cross-sectional drawings of the site.

MOUNT ETNA

On Mount Etna, Charles Lyell saw that the mountain had been built up by successive layers of lava flow. He realized that if each was perhaps a kilometre or two wide at its furthest extremity it would take up to a hundred such flows to encircle the mountain, to raise it by the depth of just one flow. It would have taken tens of thousands of eruptions to build the mountain at that rate. Yet major eruptions happen only once every few years, and Etna is over 3,300m (10,900ft) high. Even more startling, he saw between two lava layers a bed of fossilized oysters some 7m (23ft) deep. This meant that an inordinate amount of time must have passed between the two lava flows for such a thickness of fossils to form; the mountain must be far older than anyone could have conceived of. Even worse, the fossils were in some of the "youngest" rock. So how old were the layers of older rock, with all their extinct creatures that had come before? This beautifully simple reasoning led Lyell to the conclusion that the Earth must be almost unimaginably old.

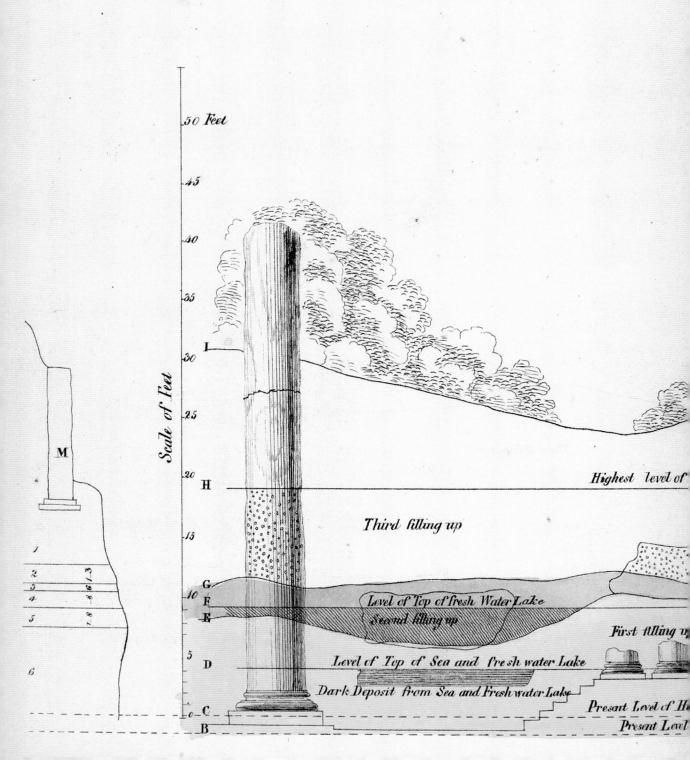

50 Feet

45

Scale of Feet

40

35

30

I

25

M

20

H   *Highest level of*

Third filling up

15

G

10

E   *Level of Top of fresh Water Lake*

E   *Second filling up*

First filling up

5

D   *Level of Top of Sea and fresh water Lake*

Dark Deposit from Sea and Fresh water Lake

C   *Present Level of H*

0

B   *Present Level*

1

2

3

4   1.3

5   1.8 - 8 - 6

6

Section of the
TEMPLE OF JUPITER SERAPIS,
Shewing the Successive Changes it has undergone.
1828

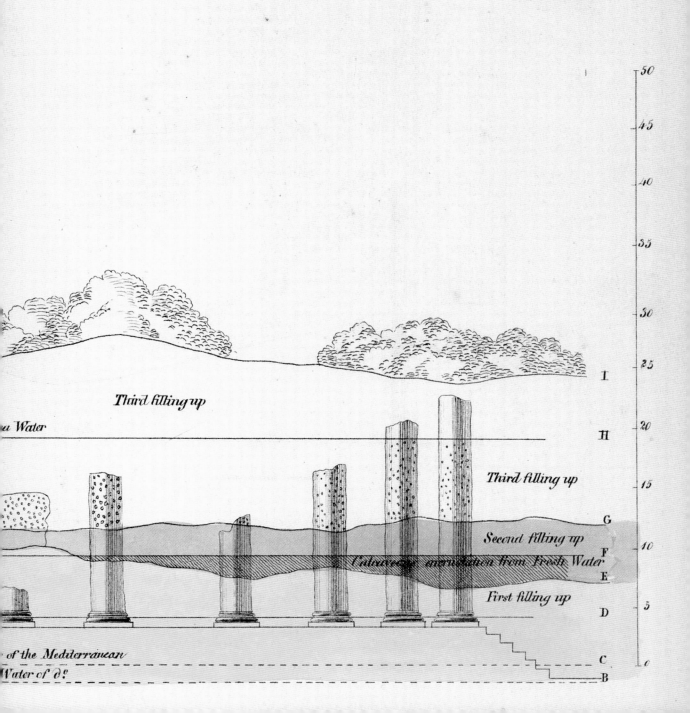

Third filling up

a Water

Third filling up

Second filling up

Calcareous incrustation from Fresh Water

First filling up

of the Mediterranean

Water of d.°

I

H

G

F

E

D

C

B

50

45

40

35

30

25

20

15

10

5

0

VESTIGES OF PROGRESS

The early years of the Victorian reign saw a staggering transformation of British society. In 20 years, over 10,000km (6,200 miles) of railway lines carved their way across the countryside, cutting journey times between cities from days to hours, and bringing cheap goods to the masses. Factory towns sprang up everywhere, bringing a new regimentation to daily life that fitted the routines of industrial production. The 19th-century concept of progress was almost palpable, as factory bosses from humble origins now became the proud owners of country estates and even took seats in parliament. And Britain itself had been transformed into the world's leading industrial power, thanks to the hard work and ingenuity of its subjects.

In this world of progress, something else was emerging: a radical new theory of life, which proposed that not only were societies capable of change, but also plants, animals, and the whole of the natural world – including mankind. In 1844 a small book rolled off the new steam-powered printing presses to become a best-seller and create a storm among Victorian society. It was called *Vestiges of the Natural History of Creation* and drew together the ideas of gradual change in geology with the puzzles of extinction in the fossil record and ever widening diversity in the natural world to propose a cosmic theory of "transmutation". It was all-encompassing, describing the formation of the Solar System according to the nebular hypothesis, whereby the law of gravity meant that smaller particles gradually accreted to form planets – not so different, in fact, from the modern accepted process. It then argued that life formed by spontaneous generation, and that the fossil record showed evidence of a gradual progression to ever more complex and superior forms of life: insects followed by fishes, followed by reptiles, then higher life forms such as mammals.

Below: *Vestiges* treated the natural world and fossil record as an example of a slow process of transmutation.

"Vestiges claimed that evolution from fish, birds, and beasts had culminated in mankind and went further in suggesting that all of this happened over the vastness of time, without the need for a God-like creator of every species and every feature of the planet."

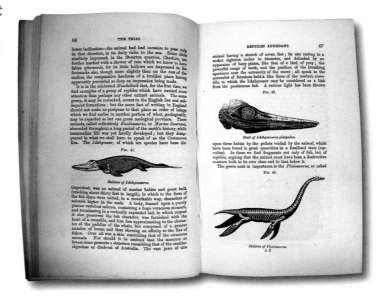

Above: Robert Chambers 1802–1871.
Chambers' outspoken views on the history of life meant that the influential publisher was widely suspected of writing the controversial *Vestiges*, though formal attribution did not come until the 12th edition of 1884.

But that was just the start of it. *Vestiges* claimed that evolution from fish, birds, and beasts had culminated in mankind (white European mankind to be specific) and went further in suggesting that all of this happened over the vastness of time, without the need for a God-like creator of every species and every feature of the planet. Rather, it allowed that God might have set out a "natural law" that thereafter progressed in an evolutionary way. The first edition sold out within days and early reviews were good. It was even read to Queen Victoria every afternoon by Prince Albert. But inevitably the writer brought a storm of criticism down upon his head, from scientists who attacked his lack of science and from religious believers who attacked his blasphemous argument that humankind had emerged from lower forms of life. It is perhaps no surprise to learn that the author was in fact "Anonymous".

It was only after his death in 1871 that *Vestiges* was confirmed to have been written by Robert Chambers, one of two brothers who ran a highly successful publishing company in Edinburgh, responsible for the world famous *Chambers Encyclopaedia*, amongst other titles. Both the brothers had been born with twelve fingers and twelve toes, and both had undergone corrective surgery as children. But while his brother's surgery was successful, Robert's left him lame, so that he turned instead to the world of books. He had become fascinated by geology in the heady intellectual years after Lyell's *Principles* was published, and saw *Vestiges* as the first real attempt to tie together natural science and the story of creation. His was perhaps the first real attempt to address the question: "how did we get here?"

LONG NECKS

Jean-Baptiste Lamarck 1744–1829

Charles Darwin, having read *Vestiges*, wrote that the author's "geology strikes me as bad and his zoology far worse". The heaps of invective that the book attracted – albeit guaranteeing its rocketing sales – were almost certainly a factor in Darwin's 20-year hesitation before going public with his own theory of evolution in 1858. Darwin was determined to ensure that his arguments were seen to be backed up by meticulously demonstrable evidence, so that he could not be accused of speculation. At the same time, he recognized that *Vestiges* had successfully opened the whole question of evolution to public debate, thus paving the way for acceptance of his theory.

Evolution was not new. Besides Chambers, Buffon years before him had suggested that animals evolved from earlier forms and Darwin's own grandfather, Erasmus, had immortalized the concept in a poem and laid it out in a huge two-volume book entitled *Zoonomia*. The ideas proposed in *Zoonomia* – that the creatures God created were designed for self-improvement – fitted well with the ethos of betterment in the whole of society. Similar ideas had been proposed and developed as a scientific concept by the French naturalist Jean-Baptiste Lamarck, once keeper of the herbarium at the Jardin du Roi in Paris. Lamarck's theory became known as the "inheritance of acquired characteristics". The most often cited example is that of the giraffe, which Lamarck argued had evolved from a creature with a shorter neck. At some point, some of the "proto-giraffes" had learned to stretch out their necks in order to reach leaves higher in the trees. Lamarck believed that somehow this facility for change was passed on to their offspring, who in turn had stretched their necks and passed their new ability on to their descendants, and so on, to bring about very gradual change. The theory became known as Lamarckism, but was widely discredited as Georges Cuvier, a believer in the immutability of species, ensured that the ideas gained no ground in French academic circles.

Lamarck at least proposed some kind of mechanism by which species would change from one form to another. That, after all, was the puzzle. If new forms of life emerged in ever more complex forms, as the fossil record showed, while others went extinct, what could possibly cause this to happen? Charles Darwin was, of course, the person who arrived at an answer, but why him, and why then did the answer emerge? Darwin came from a wealthy family and had the means to spend time indulging his interests rather than on earning money. He bolstered his position by marrying his cousin Emma, granddaughter of the rich industrialist Josiah Wedgwood. Having found medical studies in Edinburgh utterly distasteful, he transferred instead to natural history before being sent to Cambridge to train for the clergy. There he continued to focus on natural history and geology and, through his friendship with a professor of botany, found himself serving as the "naturalist" on a five-year voyage around the world. Its principal mission was to survey the coast of South America. HMS *Beagle* set sail in December of 1831. In the five remarkable years that followed, Darwin made copious notes and drawings of what he saw in the natural world around the globe. He also took with him a copy of the first volume of Lyell's *Principles*, and arranged for the second volume to be sent out to him. The concept of deep time and all that it enabled made a profound impression on him.

Left: From the 18th century onwards, the giraffe became an icon of the struggle to explain the variety of life.

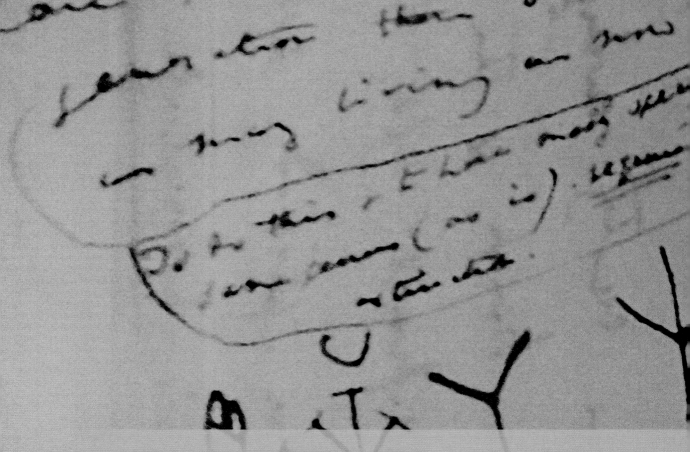

STRUGGLE FOR EXISTENCE

Charles Darwin
1809–1882

On his global journey Darwin experienced the same sense of wonder at the diversity of species that had impressed itself on Hans Sloane over a century before, and he witnessed firsthand the power of the Earth to create geological change when he experienced an earthquake in Chile. He was meticulous in his notes and careful as to the conclusions he drew from what he saw, but within two years of his return he had started to lay out his ideas on the "transformation of species". In one particular notebook from 1837 can be seen a sketch of a rudimentary tree of life, branching out from a single common ancestor – next to it, he wrote, "I think". As much as anything, his thinking was shaped by the time in which he lived. In Victorian Britain, industry and commerce thrived on competition; successful businesses made their owners rich, while those whose products were not good enough or cheap enough simply withered and failed. The apparently obvious success of Victorian capitalism was a clear indication that competition for survival was part of life. In those first years, Darwin was also influenced dramatically by reading, "for amusement", an essay by Thomas Malthus on the Principle of Population. Malthus argued that food supply could never keep pace with growing population, and that war, famine, disease, and poverty would always result, with the weakest falling into poverty and decline, and losing the struggle to survive. Politicians were using Malthus's logic to argue against support for the poorer working classes, for otherwise the British population could not be held in check. In essence, as Darwin himself later wrote, what he did was to extend Malthus's ideas of the

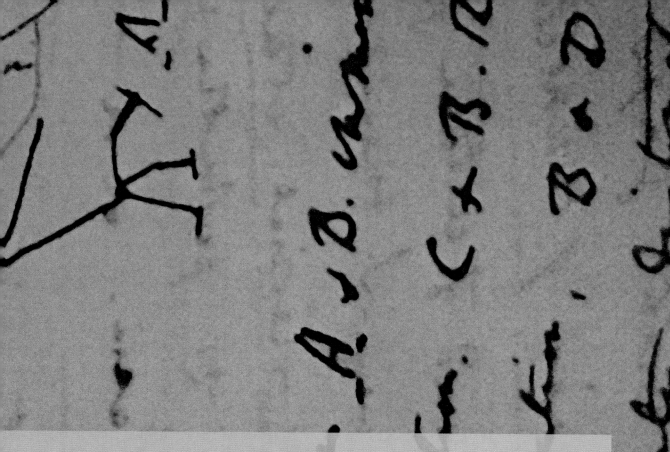

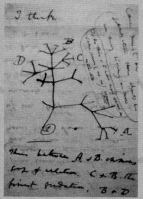

Above: Darwin's original sketch of a "tree of life". Unlike many of his supporters, Darwin did not make the mistake of creating a hierarchy among species.

limits to population growth to the natural world. He could immediately see how the possession of an advantage in the struggle for existence provided a mechanism by which species might continue to thrive, or face extinction.

With the elements of his theory in place, Darwin spent the best part of two decades amassing precise evidence to support it, and his caution was only underlined by seeing the furore that surrounded *Vestiges*. Here again, the nature of Victorian society shaped his progress. By the 1840s, natural history as a pastime had become hugely popular. As the railways began to carry more people to the leisure destinations of the seaside, so people took up the delights of studying nature. Many books were published on how to collect flora and fauna. Amateur naturalists abounded and as Darwin began to gather data he built a network of other gentlemen with whom he corresponded. Having settled with his new family at Down House, in Kent, and suffering recurring bad health, he remained a very private person, but wrote literally thousands of letters over his working life, exchanging information and samples – a practice made easier by the introduction of the Uniform Penny Post system. His network ranged far across the globe, with samples sent from countries throughout the British Empire, then still nearing the greatest extent of its global reach. Darwin carried out experiments on plants at home, studied barnacles for years and, crucially, became familiar with animal breeding. He kept and bred pigeons himself, and studied the effects that breeders have in selecting for specific characteristics in their pedigree animals, creating new varieties of dog, cat, cow, horse, plant, and pigeon.

NATURAL SELECTION

Ultimately, Darwin's theory of "evolution by natural selection" is a meticulously worked out demonstration of how nature achieves by chance, over the vastness of time, what humans have shown they can do by selective breeding – carefully choosing the creatures that they allow to reproduce. All living things compete for limited food and territory, and compete to reproduce. All offspring vary slightly from their parents. From time to time, an individual chance variation will prove advantageous in the struggle for existence; a slightly longer giraffe neck, for example, might confer an advantage in reaching more food during a lean year. Success in that struggle to survive means that the variant will produce more of its own offspring, which will themselves reproduce, so that gradually, over very, very long periods of time, the population of the successful variant will grow. More successful variation, and yet more time, may therefore lead to descendants that are so different from their ancestors that they have become a new species altogether – they could no longer breed together to produce fertile offspring.

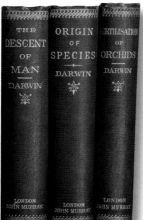

And for the varieties that do not have the advantage in the struggle to survive (the short-necked giraffes) the consequence is the reverse: their populations decline and eventually they become extinct. Above all, this whole gradual and wondrous process is enabled by the playing out of a vast amount of geological time.

Despite his comfortable world, Darwin himself was not immune to competition. In the end, he was spurred into publication of his theory only after he received a note from one of his correspondents in Malaysia, one Alfred Russel Wallace, who had independently arrived at the same mechanism of natural selection. After consultation with friends and mentors, including Lyell, Darwin and Wallace's idea was jointly announced in July of 1858. The occasion passed without notice, and over the next year Darwin hurriedly wrote up the full detail of his 25 years of work. *On the Origin of Species* was published in November 1859, becoming an immediate best-seller. It remains one of the all-time great demonstrations of a scientific argument laid out, precisely and carefully, with meticulous evidence leading inexorably to an insightful and crafted theory.

Left: *The Origin of Species* soon became an international bestseller, but foreign editions were sometimes problematic, as translators introduced their own ideas and biases.

BEYOND DARWIN

The theory of evolution by natural selection has, of course, gone on to be debated and fought over in the 150 years since its publication. There were heated rows over its implications for religious faith from the beginning, centring on the inescapable

"The answer to that question of heredity had already been mapped out by an obscure Augustinian monk, Gregor Mendel."

conclusion that humans were but variations descended from a more primitive species of ape. The concept of Darwinism in society was also adopted by a variety of political and economic visionaries. The phrase "survival of the fittest", which Darwin himself never actually coined, became used to justify the extremes of Western capitalism, the emergence of communist society, and the ideas of racial purity that underpinned the Nazi state. Most recently, the theory has been blamed for those extremes and has come under increasing attack from the religious creationist movement, particularly in the USA.

Scientifically, however, Darwin recognized that there were problems with his theory, above all that he could offer no explanation of how adaptations were passed on from one generation to another. Late in life, he even returned to the idea of Lamarckism – as did other scientists – because it seemed a good idea, and perhaps also because it fitted with the moral ideal of working hard to improve your lot, and passing on those hard-earned improvements to your children. Had Darwin known it, however, the answer to that question of heredity had already been mapped out by an obscure Augustinian monk, Gregor Mendel, who lived at the Abbey of St Thomas in what is now Brno in the Czech Republic. Mendel's painstaking work revealed the true laws of heredity: characteristics are defined by pairs of genes (although he did not use that word), one from each parent. Only one gene for each characteristic is expressed in the offspring. Some genes are "dominant" – dominant over other "recessive" genes – meaning their characteristics will appear more frequently in subsequent generations. On Mendel's death, his papers were burned by his abbot, and the work was not rediscovered until the beginning of the 20th century, but what he had effectively revealed was the process by which natural selection could operate.

MENDEL'S PEAS

Over seven years, right around the time that Darwin published *The Origin of Species*, Gregor Mendel cultivated 29,000 pea plants to test how their characteristics were passed on from generation to generation. He was trying to work out how to create better hybrid plants to produce hardier crops. At the time, most biologists argued that parents' characteristics passed down the generations by being blended in their offspring, who ended up as an average of their parents. Patently, this would result in all inherited characteristics being diluted after a few generations, and it could not explain why, for example, some children end up being much taller or shorter than both their parents, instead of halfway between the two. Mendel's painstaking work led him to conclude that there were units of heredity called "factors" (later "genes") that always operated in pairs, one from each parent. Further, he realized that genes could be either dominant or recessive, thus explaining how a particular characteristic could pass unnoticed down generations before reappearing. He developed two "laws of heredity" that are the basis for modern genetics.

CHANGING THE WORLD

But what of the question "how did we get here?" Just as geological deep time proved the key to understanding evolution, so the story of the rocks would be critical in answering that great question.

In the decades that followed the work of Lyell and Darwin, geology and biology went their very separate ways. Over the next half century biology moved ahead dramatically, with the identification of the chromosomes in the nuclei – the control centres – of cells as the place where genes are held, and ultimately by the identification of the structure of the double helix, DNA, as the chemical code of the genes themselves (See Chapter Five). Advanced molecular biology has become one of the biggest and best funded areas of science. Evolutionary theory has produced everything from enzymes that work in "biological" washing powder to bacteria that can mass produce new pharmaceuticals; it is applied at the heart of advanced computer software programmes and underpins our understanding of the progress of diseases. Today, scientists can manipulate genes and control the very process of evolution – they are even at the threshold of generating synthetic life.

Geology, meanwhile, drifted into something of a backwater. True, there were significant steps forward, particularly with the discovery of radioactivity, which enabled rocks to be accurately dated by measuring the decay of radioactive isotopes within them. Most geological effort, however, went not into debating grand theories of the Earth, or pursuing the Uniformitarian/Catastrophist debate, but into mapping and recording rocks and fossil layers in the field, and trying to get more accurate data on the relative age of strata – effectively extending the work that William Smith had been doing a century before. There were some whole-Earth theories around, but there was simply not enough evidence available to draw solid conclusions. The predominant idea around the turn of the 20th century was the notion that the planet was gradually contracting as it cooled, with the surface wrinkling up like a prune as it shrank, forming mountain ranges, valleys, ocean basins, and continents as a result. But 20th-century physicists disposed of that theory, as the discovery of radioactivity enabled them to show that the Earth's centre was barely cooling at all, and certainly not enough to account for the massive upheavals of mountain ranges like the Himalayas.

Left: Artist's impression of chromosomes in a cellular nucleus – each one contains a pair of identical DNA strands, chemically bound to various proteins in a structure called a chromatid.

CONTINENTAL JIGSAW

Then, in 1910, a little known German meteorologist called Alfred Wegener was idly browsing the pages of a new atlas when he was struck by the remarkable similarity in shape of the east coast of South America and the western coastline of Africa. He was not the first to have noticed this, but some time later he also noted the similarity in fossils of extinct animals on either side of the same Atlantic divide. This was the beginning of Wegener's theory of "continental drift", which he made public in a series of talks in 1912 and then developed over the next decade as he continued to build a career in meteorology.

A century before, Jean-Baptiste Lamarck had suggested that the continents were all steadily progressing around the globe, with the ocean currents tending to erode them on their eastern sides, while building up new sediments to the west – an idea that could be demolished as thoroughly as his theory of evolution. But Wegener's evidence for change was compelling. The more he studied the geology and natural history of the continents, the more he saw signs of them having once been joined together and moved apart: the remains of glaciers and mountain ranges that lined up, fossils of tropical plants found in today's Arctic, and similarities in populations of living animals and in layers of rock strata across the globe. Like a jigsaw, he worked out a pattern for how they could all have fitted together, and arrived at a map of a supercontinent that had once dominated the surface of the planet. He named it "Pangea".

The geological community ridiculed the idea – how could huge slabs of continental crust possibly plough their way through the rocks of the ocean basin? Wegener's evidence was all circumstantial, they said. In 1930, he set off on an expedition to monitor Arctic weather and also, he hoped, to obtain measurements that would prove that the landmasses of Greenland and Scandinavia were moving rapidly apart. At the height of the Arctic winter, on his 50th birthday, Wegener tried to return to his base camp, across the Greenland ice cap, for more supplies. He never arrived – his frozen body was found the following year. Even after his death his theory continued to be vigorously attacked, and young geology students had it made known to them that enquiring about continental drift was not going to do their academic careers any good.

"The geological community ridiculed the idea – how could huge slabs of continental crust possibly plough their way through the rocks of the ocean basin?"

A series of global maps shows the tectonic movement of the continents from the breakup of Wegener's "Pangea" to the present day.

Jurassic Period – c.195 million years ago

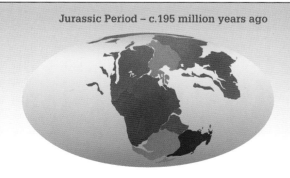

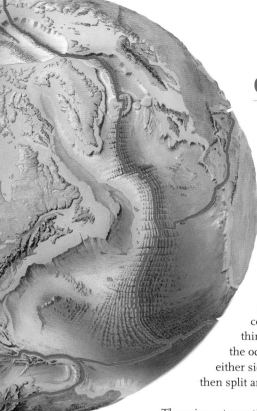

COLD WAR CLUES

The hard evidence that had eluded Wegener came eventually from a very different direction. The dramatic growth of submarine warfare during World War II and in the early years of the Cold War led to an urgent need to understand the ocean floor, to map what had been until then a hidden world. Resources were poured into the new science of oceanography, and the results were remarkable. Three astonishing new pieces of evidence emerged. First, it became clear that there was a huge undersea mountain range running the entire length of the Atlantic Ocean like a spine along the centre; it actually reaches right up above the surface to form the island of Iceland. Then, it became apparent that the rocks of the ocean floor, which geologists had always assumed were the most ancient, with the continental crust being laid on top, were in fact very young. And thirdly, the new science of paleomagnetism revealed that the rocks of the ocean floor had a pattern of magnetism that was exactly mirrored on either side of the ridge, suggesting that rock once formed in the centre and then split and moved apart in opposite directions.

Above: A globe of the Earth revealing the underlying structure of continental plates, with boundaries concentrated along continental margins and at deep ocean ridges.

These incontrovertible bits of hard data were all assembled in the mid 1960s as the Theory of Plate Tectonics, which is now accepted as the mechanism by which the Earth's crust is formed as molten volcanic rock at mid-ocean ridges. These ridges form a vast interlinked network around the globe. According to plate tectonics, the sea floor spreads apart at the ridges, pushing the continental crust along with it. Elsewhere, slabs of crust collide with each other, with one being pushed beneath the other, disappearing down into the molten rock below. The surface of the planet is divided up into a myriad of these "plates", like the cracked surface of a boiled egg. Where they grind and crash into each other, they form fault lines, such as the San Andreas Fault in California, creating earthquakes and volcanic eruptions, or forcing mountains to rise. The mechanism driving all this is convection – molten rock rising and cooling rock falling – in the warm mantle beneath the crust.

Cretaceous Period – c.125 million years ago	Paleogene Period – c.60 million years ago
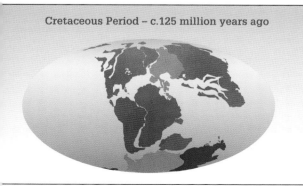	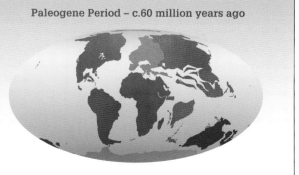

LIFE AND DEATH

The science of geology was revitalized by the discovery of plate tectonics. Suddenly all sorts of geological mysteries, such as why earthquakes and volcanoes happen where they do and why mountain ranges are pushed up, made sense. There are also huge practical benefits, as geologists can now more accurately predict where oil-bearing rocks will be found, and direct the mining of minerals that have been formed in the depths of the Earth and pushed to near its surface. But tectonics also has revealed that there is an inextricable link between geology and life. We now understand that plate tectonics is key to the movement of carbon throughout the Earth, with carbon dioxide being belched into the atmosphere by volcanoes, then drawn out of the air by plants and trees, which feed animals that die and decay to form layers of sediment on the ocean floor, which eventually fall back into the molten planet to begin the cycle again. Tectonics explains how and where layers of dead animals and plants have, over millions or billions of years, formed layers of coal or oil in ancient ocean basins. The theory is crucial to the intricate balance between oxygen and carbon dioxide in the atmosphere. In many ways, the story of life on Earth is the story of geology itself.

As for the direction that the evolution of life has taken, it too has been profoundly influenced by the movement of the planet. As the continents worked their way through different shapes at different latitudes, so the species that were evolving on them or in the oceans around them found themselves having to adapt to different climates, or compete with other species that were able to move into their territory. Some flourished, while others became extinct, creating opportunities for new forms of life to emerge. The rising up of huge mountain ranges like the Himalayas – which occurred as the plate carrying the Indian subcontinent crashed into the one carrying Asia – created barriers to migration or changed the weather system in the upper atmosphere, the results of which were new environments, floods, or droughts for life to contend with. The shifting position of the continents on the globe also changed the circulation of the oceans, placing new stresses on marine life and triggering the rapid onset of ice ages or their sudden thawing. From time to time, huge life-threatening events occurred in the form of vast earthquakes, tidal waves from bursting glacial lakes, or volcanic eruptions, which sometimes flowed continuously for thousands of years at a time. Add to all this the greater violence of the early Solar System, where asteroid and comet impacts were more frequent, and we can see the traces of true catastrophes on our planet. The fossil record and the geological evidence reveal that there have been five mass extinctions that have annihilated more than half of the species living at the time, one of which came close to wiping out all life on Earth. The comet impact that struck the planet 65 million years ago, marking the end of the era of the dinosaurs, is only the most famous, not the largest.

Right: A satellite view of volcanic activity and glaciers on Iceland, one of the few places in the world where an oceanic ridge, marking the separation of two tectonic plates, runs above ground.

CONTINENTAL CREATION

Earth-orbiting satellites allow us to see the entire ocean floor. What stands out is the extraordinary network of mid-ocean ridges that run like scars through the centre of every one of the oceans. Together these make up one continuous mountain range of almost 60,000km (37,000 miles) in length and form the highest mountains on the planet, despite being almost entirely hidden from our view. Small islands such as Iceland, the Azores, Bermuda, Ascension, and Tristan da Cunha are really mountain peaks, marking the points where the ridge breaks through the surface of the Atlantic. The mid-Atlantic ridge runs all the way from the Arctic to the Antarctic oceans – the longest mountain range in the world. The ridges form one set of boundaries of the Earth's tectonic plates. They mark the point at which new ocean floor is being formed, spreading out and driving the colossal movement of the Earth's crust. The Atlantic Ocean, for example, is less than 180 million years old, and only began to form when the supercontinent Pangea started to break apart. The world's landmasses and oceans only reached their recognizable configuration some 50 million years ago.

LUCKY SURVIVORS

So the answer to the question "how did we get here?" turns out to be quite clear. We arrived by chance. This lump of molten rock has been responsible for the appearance – and disappearance – of every species that has ever lived, including us. We are the product of life's response to an ever changing and violent planet, not the pinnacle of a great chain of being. Our line is that of the survivors, the ones whose ancestors succeeded in the struggle for existence while others around them failed. A chance climate change here, an earthquake there, a drying ocean or a slide into an ice age, and the story may have been very different. If we were to turn the clock back to those eons ago when life began, and let evolution and the planet take their course again, then there is little certainty that we would be the outcome.

We are extremely lucky to be here. It is a sobering thought.

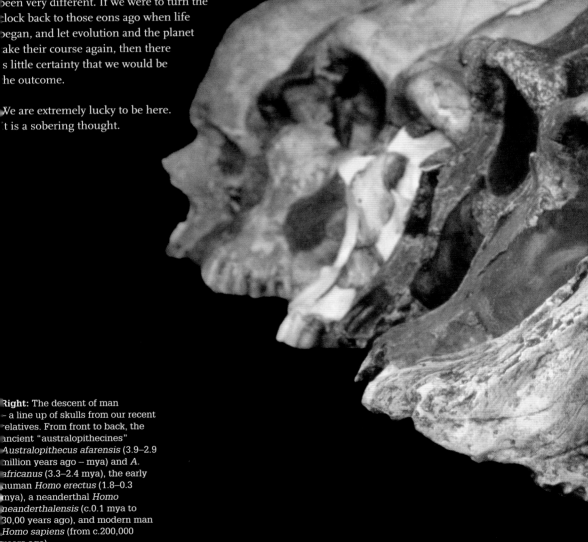

Right: The descent of man – a line up of skulls from our recent relatives. From front to back, the ancient "australopithecines" *Australopithecus afarensis* (3.9–2.9 million years ago – mya) and *A. africanus* (3.3–2.4 mya), the early human *Homo erectus* (1.8–0.3 mya), a neanderthal *Homo neanderthalensis* (c.0.1 mya to 30,00 years ago), and modern man *Homo sapiens* (from c.200,000 years ago).

Power

CAN WE HAVE UNLIMITED POWER?

In the last thousand years, human ingenuity has released vast quantities of energy from nature and created a world shaped around the on–off switch. The invention, design, and building of new machines and the discovery of successive new sources of energy are the backbone of this story. But it is a story of two threads that are intertwined: one tells of how we learned what power does; the other of how we discovered what power really is. Yet the search for new energy sources was led not by theorists trying to un-cover grand scientific laws, but nearly always by practical people on the make: the inventors, industrialists, and investors who saw how to exploit energy, and who dreamed the dream of unlimited power, and fortune. Attempts to reach that goal have not only helped to create the modern world but, when theory finally caught up with practice, also given us some of the most profound insights in the history of science – about the nature of time, and the fate of our Universe.

Left: Electrical power has enabled human beings to finally conquer the night, and created our modern 24-hour civilization – but the path to it has also revealed fundamental principles of physics and laws of nature.

WATER WORLD

It begins with water. The use of waterwheels dates back to the ancient civilizations of Mesopotamia; Babylonian and Sumerian inscriptions refer to them, but do not make clear how they were used, and no description remains. Han Dynasty China had waterwheels in operation from the 3rd century BC, using them to power furnace bellows and trip hammers for forges. But by the time of the Ancient Greeks and Romans water had been harnessed to power mills on a huge scale. At Barbegal, near Arles in southern France, there still stand the remains of what has been called "the greatest known concentration of mechanical power in the ancient world". Here, an aqueduct had been built to supply water from the Alpilles mountains to the Roman city at Arles, and along its path the water flowed over two rows of huge waterwheels, 16 in all, powering 16 mills cut into the hillside. The water was channelled down the slope, with the outflow from one wheel pouring over the next one below. It has been estimated that the mills turned out 4.5 tonnes of flour a day for the people of Arles and its surrounding area.

Much of Muslim Spain, North Africa, and the Middle East was powered by horizontal waterwheels as well as vertical ones from the 7th century, and Islamic engineers are credited with the invention of the flywheel – the concept of a faster spinning wheel whose weight smoothes out any unevenness of the main wheel if the water supply changes. Crankshafts, gears, water turbines, and dams to contain the water supply were all features of large industrial complexes in Muslim Spain, with cloth mills, paper mills, and even steel mills powered by water. In medieval Europe, too, from the 8th to the 15th centuries, there were malt mills, tanning mills, and hemp mills, mills for forging iron, ore crushing, sharpening tools, and sawing wood. It was a water-powered world, all born of the ingenuity of practical mechanics and engineers, and three thousand years of trial and error.

Top: The remains of a giant medieval Islamic waterwheel preserved at Hama in Syria.

Above: Ruins of the great aqueduct at Barbegal in southern France – an ancient power line to the watermills.

Below: A view of "norias" on the Orontes River at Hama – another type of waterwheel used to lift water into an aqueduct.

DUTCH WIND

Simon Stevin
1548–1620

The attempt to link both theory and practice in the exploitation of power only really arrived with the work of a remarkable Dutch mathematician and inventor called Simon Stevin. When it came to natural power supplies, the 16th-century Netherlands – the "Low Countries" comprising most of today's Belgium, Netherlands, and Luxembourg – had a disadvantage. Their name encapsulates the problem: the land is substantially flat, and there was little naturally fast-flowing water to power the mills. Instead, the landscape of the flat coastal region along the North Sea provided as consistent a supply of wind as could be expected anywhere in the developed world at the time.

Of course, wind had long before been harnessed to drive sailing ships. Windmills, however, which originally operated on a vertical rotating axis, had emerged in Persia in the 9th and 10th centuries and spread throughout Europe. But it was in the Low Countries that their use was developed to power an empire. In the mid 16th century, the Netherlands were part of the Hapsburg Empire, ruled by the Holy Roman Emperor, Charles V, and then by his son, Philip II of Spain. The half century that followed saw a long rebellion waged against Spanish rule resulting in the division of the lands into the areas approximating today's Belgium and Netherlands. The division also saw a massive shift of wealth from Antwerp, which had been the dominant trading city of northern Europe, when the Protestant half of its population fled north towards Amsterdam. The shift marked the beginning of an explosive period of economic growth for the Dutch United Provinces, and Holland became the most powerful trading nation in the world.

"The shift marked the beginning of an explosive period of economic growth and Holland became the most powerful trading nation in the world."

The background to all this turmoil was a revolution in the generation of power. Simon Stevin, an engineer and architect born in the Belgian town of Bruges, was involved in the design of critical military fortifications in the fight against Spain. He was also granted an astonishing number of patents by the States of Holland, and it is from these, and from records of disputes that he had over the commercial exploitation of his ideas, that it is possible to build a picture of the engineering work that he was undertaking. He worked on sluices, dredging machines, and mills. The low-lying boggy land of the Netherlands suffered repeatedly from flooding in storms that howled off the North Sea, and the Dutch had for generations protected their land by building dykes to keep out the ocean, using windmills to drain the low-lying areas and to pump out flooded waterways and harbours. What made Stevin's ideas special was that he was schooled as a mathematician and endeavoured to apply theory and calculation to the building of more efficient wind power. His work was a mixture of the practical and theoretical: calculating the ideal angle to fit the scoop-wheel paddles to lift the most water, while fitting leather flaps on the paddles to minimize the spillage of water as it was lifted; and also calculating the most efficient speed of rotation, and the ideal size of the cogs in the gearing.

His major contribution to the Dutch nation, however, was to significantly improve the ability to reclaim land by arranging windmills in stepped rows, so that water could be pumped out from deeper and deeper low-lying areas. As the land got lower and the dykes got bigger, new layers of stepped windmills had to be built to preserve the newly created land. At its deepest, the coastal land was maintained by a series of up to 14 stepped windmills. It is argued that the "Golden Age" of the Dutch mercantile empire could not have happened without the successful drainage of land to underpin an agriculture large enough to support the population. By the mid 17th century the Dutch East India company, founded in 1602, was the richest private company the world had ever seen – the first multinational corporation. Amsterdam and Delft were overflowing with porcelain from China, as well as spices, cocoa, and rice, all of which needed grinding – in mills. Windmills were used to produce textiles, paper, oil, spices,

"Amsterdam and Delft were overflowing with porcelain from China, as well as spices, cocoa, and rice, all of which needed grinding – in mills."

and flour. In 1594, a Dutchman named Cornelis Corneliszoon had patented the wind- powered sawmill, using a crankshaft to convert the windmill's rotary power to the back and forth action of a cutting blade. The result was a rapid expansion of the shipbuilding industry in Holland, not only for its own great trading fleets, but also for constructing the navies of other European countries. By 1600, half of the British naval fleet was built by the Dutch.

The 17th-century Dutch Golden Age established what is now recognized as the first modern economy, with insurance, retirement funds, a central bank, and the first full-time stock exchange, in Amsterdam. There was even the first boom and bust cycle of inflation, based on absurdly high prices for tulip bulbs. Introduced to Europe from the Ottoman Turks in the late 1500s, serious cultivation of the tulip began after a botanist from the University of Leiden imported a variety that thrived in the colder Dutch climate. The flower became a must-have status symbol, with more and more varieties and colours being bred. Prices for the latest trends in bulbs reached an extraordinary high in February 1637; it is claimed that a single bulb was traded for 5 hectares (12 acres) of land. But within three months the market had collapsed, and many growers and traders were ruined. The first modern economy had experienced the bursting of the first investment "bubble".

The Dutch had the highest standard of living in the known world of the 1600s. Their economic revolution had been built on the systematic exploitation of energy from the air. And they continued to rely on the wind well into the 19th century. At the same time as steam was beginning to transform British society on the other side of the North Sea, there were over ten thousand windmills powering industry and commerce in Holland.

Simon Stevin was interested in other ways of exploiting energy. He designed and built a wind-powered car that he used to transport his friend Prince Maurice of Nassau (who went on to lead the successful revolt of the States of Holland against Spain) and 26 others in a joy ride along the beach at Scheveningen, claiming to have travelled faster than the speed of horses. His interest in numbers led him to introduce the use of double entry bookkeeping into State accounting, and he also is credited with inventing the concept of the decimal system to replace fractions.

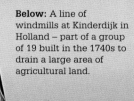

Below: A line of windmills at Kinderdijk in Holland – part of a group of 19 built in the 1740s to drain a large area of agricultural land.

PERPETUAL MOTION

The notion of perfect power did not pass Stevin by. The dream of free energy had beguiled some of the best minds that had come before him. The "magic wheel" invented in Bavaria used lodestones (magnets) attached to a ferris wheel to make the wheel spin apparently by itself – until stopped by friction. The French medieval architect and mason, Villard de Honnecourt, who recorded the building of the great medieval cathedrals in his sketchbook, left a design for a perpetual motion machine based on weights attached to a wheel. Even Leonardo da Vinci, amidst his designs for helicopters, submarines, and parachutes, also left behind a plan for an overshot waterwheel that he claimed would provide energy in perpetuity. Stevin, however, became intrigued by a particular theoretical device involving not wheels but balls – a chain of rotating balls that ran down an inclined plane and then up a shorter one. It was claimed that because more balls were on the "down" side than the "up" side, the chain would run forever. Stevin was able to calculate that, sadly, it would not, and in so doing established a principle of static equilibrium in mechanics that still holds good today.

At around the same time that Galileo was famously experimenting with projectiles (see Chapter One), Stevin was doing his own analysis of force, mathematically dividing action on a slope into horizontal and vertical components. He was a polymath, a man interested in understanding principles and testing evidence, and one of the earliest "scientific" thinkers. One of his goals was to recover the lost wisdom of the past and to usher in a new age of universal knowledge which, by comparing translations of words across different European languages, he calculated would be best communicated in Dutch. In this way, he believed it could reach beyond the Latin-speaking elite. In the front of his work on decimal calculations he wrote the following dedication: "Simon Stevin wishes the stargazers, surveyors, carpet measurers, body measurers in general, coin measurers and tradespeople good luck." Unfortunately, Dutch did not become the language of science, and this is perhaps one of the reasons that despite the remarkable science he undertook, published far ahead of any others, so few people ever heard of him, either then, or now.

Stevin's work represents a rare moment in the early history of power when theory really did influence practical outcomes. But what he and no one else knew then was that perpetual motion contravenes the first and second laws of thermodynamics; it would be another two hundred years before theoreticians worked out that such laws existed, and in that time the world of power would be transformed by people prepared to get their hands very dirty and to struggle for every penny they could get.

"Stevin's work represents a rare moment in the early history of power when theory really did influence practical outcomes."

Right: A series of sketches by Leonardo da Vinci showing a variety of waterwheels and Archimedes screws (devices for raising water), annotated with Leonardo's unmistakeable "mirror writing".

Below: Typical designs for perpetual motion devices often involved waterwheels linked to channels that caused the water released at one point to return and drive the wheel at another point.

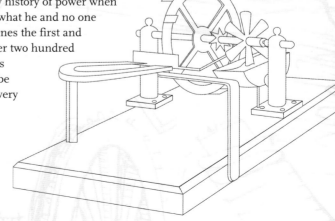

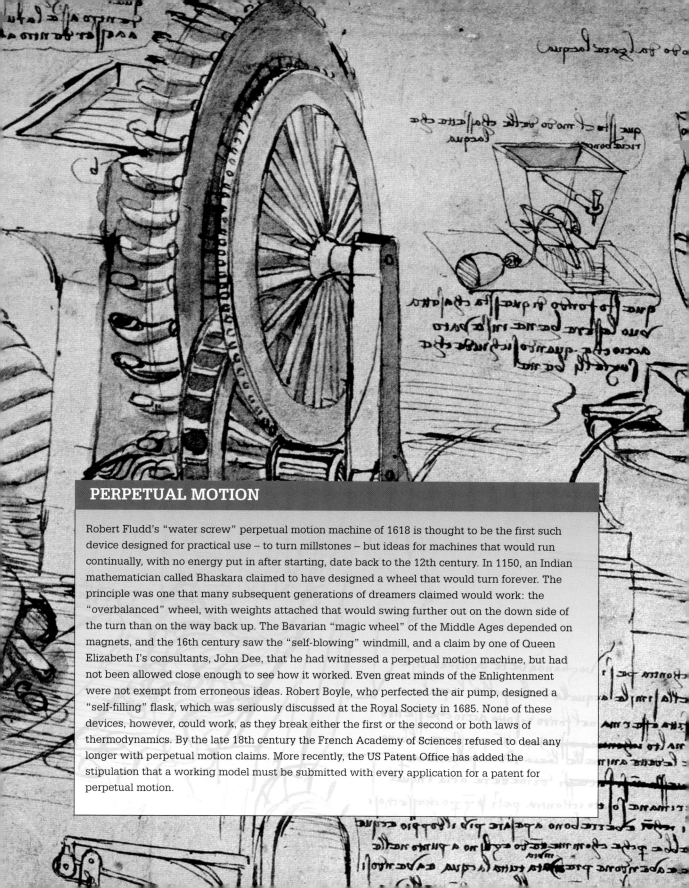

PERPETUAL MOTION

Robert Fludd's "water screw" perpetual motion machine of 1618 is thought to be the first such device designed for practical use – to turn millstones – but ideas for machines that would run continually, with no energy put in after starting, date back to the 12th century. In 1150, an Indian mathematician called Bhaskara claimed to have designed a wheel that would turn forever. The principle was one that many subsequent generations of dreamers claimed would work: the "overbalanced" wheel, with weights attached that would swing further out on the down side of the turn than on the way back up. The Bavarian "magic wheel" of the Middle Ages depended on magnets, and the 16th century saw the "self-blowing" windmill, and a claim by one of Queen Elizabeth I's consultants, John Dee, that he had witnessed a perpetual motion machine, but had not been allowed close enough to see how it worked. Even great minds of the Enlightenment were not exempt from erroneous ideas. Robert Boyle, who perfected the air pump, designed a "self-filling" flask, which was seriously discussed at the Royal Society in 1685. None of these devices, however, could work, as they break either the first or the second or both laws of thermodynamics. By the late 18th century the French Academy of Sciences refused to deal any longer with perpetual motion claims. More recently, the US Patent Office has added the stipulation that a working model must be submitted with every application for a patent for perpetual motion.

Left: While members of the Lunar Society chose the night of the full Moon for their meetings because of the additional light it offered, they still cheerfully referred to themselves as "lunaticks".

THE LUNAR MEN

A full Moon in the 1770s meant a bad night's work for the footpads and robbers who might waylay a gentleman on his way home from the coffee house or dinner at a friend's. Around the rapidly growing industrial town of Birmingham, in the midlands of England, it marked a monthly appointment eagerly anticipated by an unusual group of friends who chose the date in order to rely on the light of the Moon to see them home. These were the members of what became known as the Lunar Society. They were an assorted bunch. Clubs and societies were all the rage at that time, but this was no hell-fire gathering or drinking party for the aristocracy. Its "members" were all practical people who worked for a living and were concerned with advancing the transformation of the society they lived in. Some were nonconformists, others were established churchmen; some were radicals, others were traditional in their views. But they left their politics and their religious beliefs behind them at the door, because what they talked about was the thrilling new scientific and technical knowledge that their remarkable century was generating. Their names read like a roll call of the early Industrial Revolution: Josiah Wedgwood, the potter who introduced the factory system; Erasmus Darwin, grandfather of Charles Darwin (see Chapter Three); James Watt, of steam engine fame; Matthew Boulton, entrepreneur and industrialist; John Wilkinson, the ironmaster; Richard Arkwright, the man who mechanized the British textile industry; and occasionally, visiting from America, Benjamin Franklin, libertarian and political revolutionary.

> "They talked about the thrilling new scientific and technical knowledge that their remarkable century was generating."

Above: A page of balloon designs from a 1785 edition of the *Gentleman's Magazine*.

Their interest was in ideas, discoveries, and invention, and what you could do with them. The 18th century saw a stream of new ideas, appearing in publications as wide ranging as the *Philosophical Transactions of the Royal Society* to the *Gentleman's Magazine*. It was the century that saw an end to John Harrison's long struggle to perfect the marine chronometer, solving the problem of measuring longitude at sea. It was the century that saw the breaking down of the Ancient Greek "elements" of air and water into their constituent chemistry (see Chapter Two); the discovery of a new planet, Uranus; and the astonishing collections of plants and animals brought back from the great ocean voyages of navigators such as Captain James Cook. It was the century that saw the emergence of the "Age of Reason", when writers and philosophers began to argue that enlightened rational thinking would banish mystical belief and superstition, and usher in a new era of political and intellectual freedom, along with material progress, all driven by the two "free nations" of Holland and England. It became known as the Enlightenment.

ELECTRICKERY

More than anything, the scientific phenomenon that characterized the era of the Enlightenment was electricity. The name "electricity" was coined by William Gilbert, physician to Queen Elizabeth I, who had spent years studying magnetism and concluded that the Earth itself was magnetic, thus explaining why a compass needle would always point north. His book of 1600, *De Magnete*, also included details of his experiments to create static electrical charge – although he didn't know what it was – by rubbing a cloth on amber and other materials. For decades, the favoured method of creating static was to rub a glass rod vigorously against a cat's fur. The result would enable the rod to attract feathers or threads or scraps of paper, or to be discharged with a small spark. Some time around the 1660s, a German called Otto von Guericke, better known for his work on vacuums, invented a "friction machine" to generate electrical charge. It consisted of a rotating globe of sulphur, against which a hand could be rubbed.

The newly formed Royal Society at the turn of the 18th century was a place in London where natural philosophers (the name "scientist" came over a century later) met to discuss, theorize, and experiment, and was part of a network that spread right across Europe, with intellectual links to similar scientific academies in Florence, Paris, and Berlin. Isaac Newton was president, and his apprentice and "prize experimenter" was Francis Hauksbee. Hauksbee found that if he used a glass globe instead of Guericke's sulphur ball he could consistently generate a charge that could really be worked with. After he published *Physio-Mechanical Experiments* in 1709, electrical demonstrations took off in lecture halls, private parties, and public shows. A former cloth-dyer, Stephen Gray, took up electricity with a passion, and began systematically experimenting with it, perfecting techniques for passing charge along very long lengths of thread and wire, and identifying the different materials that fell into the categories of insulators and conductors of electric charge. One of Gray's most impressive demonstrations was to suspend a small boy from the ceiling on silk threads, charge up his feet with the friction machine, and then let the boy attract pieces of metal chaff with the end of his finger. Spark generators small and large were all the rage, and dinner party guests were treated to cutlery that sparked or even electric kisses from their hostess.

Scientifically, it was only a slightly different story. In the mid 1740s, a Dutch doctor of medicine called Pieter van Musschenbroek was professor of mathematics at Leiden

> *"The name 'electricity' was coined by William Gilbert, who spent years studying magnetism and concluded that the Earth itself was magnetic."*

Right: It was Enlightenment thought that first connected these massive blasts of lightning with the tiny sparks of static electricity.

Left: An early Musschenbroek capacitor or "Leyden jar". The ability to "store" electricity for use in experiments was key to a growing understanding of this curious phenomenon.

University, where he invented the world's first capacitor – a large glass jar, part filled with water, with a wire passing through the stopper at the top. Musschenbroek discovered that the jar could be charged up using a friction machine, and would then store the charge until it was released at will, producing a massive electric shock. He wrote about his "new but terrible experiment, which I advise you never try yourself", and his negative marketing technique ensured that the device was adopted by natural philosophers and showmen across Europe. The device was perfected and soon batteries of several jars in a row were being used to deliver massive discharges – enough to do serious harm, or even kill a small animal.

Stored electricity was an opportunity for serious study, but the natural philosophers found entertaining spectacle much too tempting. In France, the priest and natural philosopher Abbé Nollet gave Musschenbroek's device the name "Leyden Jar" and used one to electrify a row of 200 Carthusian monks holding hands, watching them all jump as the charge passed down the line. In America, Benjamin Franklin promoted his reputation as an enlightened thinker through his famous experiment to draw charge from a lightning bolt (not missing the opportunity to promote his patented lightning conductor along the way). But even he could not resist offering his guests electrified champagne or feeding them turkey that had been killed with an electric shock.

Through it all, however, one thing became very clear. No one had a clue what electricity really was. William Gilbert had called it an effluvium; to Hauksbee it was electrical force. Newton thought it might be part of the principle of life itself; Joseph

Above left: An illustration of one of Stephen Gray's early experiments into transmission of static charges.

Above right: William Gilbert's 1600 study of magnetism, *De Magnete*, influenced the development of experimental science in the 17th and 18th centuries.

Priestley thought it was merely a fluid that regulated life. Franklin, meanwhile, argued for a two-fluid theory. For a while it became known as electrical "vertue". John Wesley, the founder of the Methodists, believed that, whatever it was, it could be a universal cure for the poor of society and he toured the country holding daily surgeries to treat

> ### "Through it all, however, one thing became very clear. No one had a clue what electricity really was."

his followers with electric charge. And of course, whatever else it was, electricity was an opportunity for making money. The showmen made money from their entertainment and, at a time when medicine was a business like any other and had little better to offer, the obvious effect of electricity on the human body meant

that it rapidly became seen as a medical panacea. Perhaps the extreme example is the case of a failed Edinburgh medical student who set up the Temple of Health, featuring an "Electrical Celestial Bed", to help infertile couples conceive a child. For a while, Emma Hamilton, later to become famous as the mistress of Britain's most famous admiral, Lord Nelson, worked there as a dancer and "vestal virgin".

THE VACUUM

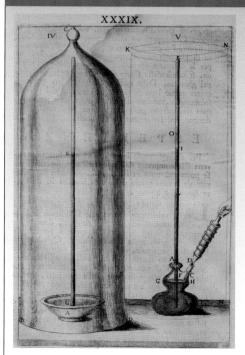

XXXIX.

The notion that "nature abhors a vacuum" probably stems from Aristotle, who argued that empty space would always try to draw something into itself. The concept was the subject of heated debate until the middle of the 17th century, when Evangelista Torricelli, who succeeded Galileo as professor of mathematics at the University of Pisa, was investigating why it was not possible to raise water higher than approximately 10m (33ft) using a suction pump. (We now know it is because it was not possible to overcome the pressure of the atmosphere.) He filled a tall glass tube with mercury, placed it upside down in a bowl of mercury and watched the height of the column of mercury fall until it was supported by the pressure of the atmosphere acting down on the liquid in the bowl. Torricelli declared that the space left behind at the top of the glass tube must be a void – a vacuum. He had invented the first barometer, because the height of the mercury column varied with atmospheric pressure. In 1650, the German Otto von Guericke built the first vacuum pump, which became refined by others until it was an essential piece of scientific equipment, enabling the study of gases.

STEAM

It was in this heady atmosphere of knowledge, ignorance, wonder, and profit that the men of the Lunar Society maintained their friendships. The two key players were Erasmus Darwin, who was forever proposing new ideas and inventions, such as a steam-powered car, a horizontal windmill, or a robotic talking head, and Matthew Boulton, who always had his eye most firmly on the money. While Darwin wondered if electricity was key to the human soul, Boulton saw it very firmly as something material. But both believed in the ideal of progress, and both believed they could help achieve it. Boulton inherited a business making small metal luxury goods: buttons, buckles, candlesticks, snuff boxes, tweezers, and the like. It was known as toy manufacture. He was a good businessman with an eye for marketing and patronage, successfully raising his profile in London with the royal family. He built the business up and moved into an expanded factory site with a large house at Soho, just outside Birmingham, which became a byword for industrial innovation and success. Soho had a very uneven water supply, which made its waterwheel unreliable, and Boulton's thoughts turned increasingly to steam power. Against this background, the network of Lunar Society friends and correspondences introduced him to a man together with whom he would transform the industrial world: James Watt.

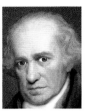

James Watt
1736–1819

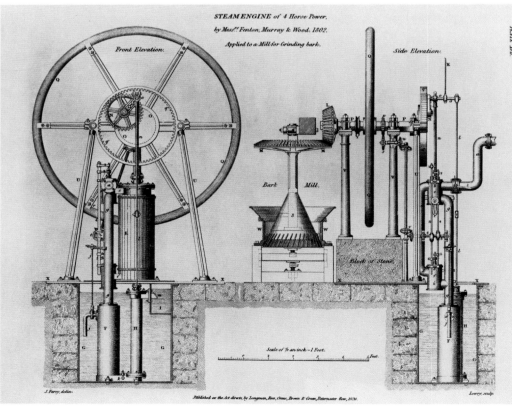

STEAM ENGINE of 4 Horse Power,
by Messrs. Fenton, Murray & Wood. 1802.
Applied to a Mill for Grinding bark.

Front Elevation.

Side Elevation.

Bark Mill.

Block of Stone.

Scale of ½ an inch = 1 Foot.

Published as the Act directs, by Longman, Rees, Orme, Brown & Green, Paternoster Row, 1826.

Left: Illustration of an early steam engine by Murray, Fenton, and Wood. The series of wheels above the piston on the left-hand side form a "hypocycloidal gear", devised in order to avoid patent restrictions.

Watt, a Scottish instrument maker, owned a small shop in Glasgow where he repaired and sold everything from bagpipes to astronomical quadrants and devices for perspective drawing. A popular myth is that Watt took inspiration for building the steam engine from watching a bobbing kettle lid as a small boy; in fact, it was not Watt who invented the steam engine at all. And the real story is a much messier, dirtier, and more significant one.

WATT'S INVENTION

In the Newcomen "atmospheric" engine, steam from a boiler is piped into a large metal cylinder to fill the space beneath a piston. Then a jet of cold water is piped in, causing the steam to condense rapidly to water. This creates a vacuum inside the cylinder, and the pressure of the atmosphere outside now forces the piston hard down into the cylinder, pulling down a beam, which drives a pump at the other end. Then the weight of the beam tilts it back down, lifting the piston at the other end, which draws in more steam, and the cycle starts over again. James Watt's key invention was the "separate condenser", into which the steam was piped in order to cool it down, leaving behind the vacuum in the main cylinder. Thus energy was not wasted in reheating the cylinder. Watt's steam engine used 75 per cent less coal than a Newcomen engine, making it an economic necessity for mine owners.

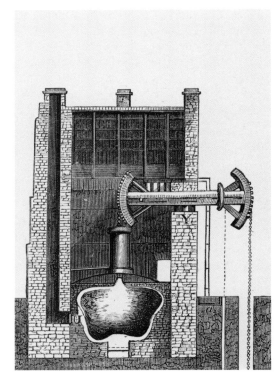

Steam engines had been around for quite some time by the Age of Enlightenment. There is a record from the 1st century AD of an "aeolipile", a little ball that spins round driven by jets of steam squirting out, and there are designs for steam turbines from the Ottoman Empire and Italy in the 16th and early 17th centuries. An English inventor called Thomas Savery patented "An Engine to Raise Water by Fire" in 1698, which he named the Miner's Friend. Then there was the "steam digester", a rudimentary pressure cooker built by Denis Papin, from France. But the first practical steam engine was the "atmospheric engine" developed by Thomas Newcomen, a Baptist lay preacher and ironmonger from Devon who took over the Savery patent, radically improved the idea, and went on to build some 75 machines that were installed throughout Britain. Mining in Britain was under pressure to go deeper and further to extract more and more tin, copper, lead, and coal for industry, and that meant a growing need for ways to pump water out of deep mine shafts. It was this demand for pumping that drove the development of steam power.

Newcomen's engine was spectacularly inefficient, because it required the main cylinder to be cooled down to condense steam and create a vacuum for each stroke of the piston. Thus a huge amount of energy was wasted in reheating the main cylinder again. This was the world of steam when James Watt first became interested in it. Prompted by a friend at Glasgow University, the two of them wondered if it might be possible to make a steam-powered car. Watt tried to build a model, but it did not work.

Above: An illustration of a full-scale Newcomen engine alongside a model of the device built for Griff colliery in Warwickshire.

In 1760, the University asked him to repair a model of a Newcomen engine. He did his best, but was unimpressed with its poor efficiency. Watt was a depressive, but a practical engineer and a perfectionist; and he enjoyed puzzles. No one asked him to design a better steam engine, but from that moment he set about trying to do so. His great insight, which he later wrote had come to him in a flash of inspiration while walking across Glasgow Green, was simple but clever. He realized that if the steam could be cooled and condensed in a vessel separate from the cylinder containing the piston, then the cylinder would remain hot the whole time, thus making the engine far more efficient in its use of heat and fuel. So he did not invent the steam engine; he invented the separate condenser. But it has to be said that it was a spectacularly successful modification.

"Mining in Britain was under pressure to go deeper and further to extract more and more tin, copper, lead, and coal."

THE FIRST STEAM CAR

James Watt's attempts to make a steam-powered vehicle were a failure. So spare a thought for Nicholas-Joseph Cugnot, a French military engineer from Lorraine. In 1769, he successfully built a model "fardier à vapeur", a three-wheeled, piston-driven vehicle powered by an atmospheric steam engine. A full-size machine weighing 2.5 tonnes was tested the following year with a view to being adopted by the French army for drawing cannon to the battlefield. It succeeded in carrying four people at a speed of 4km (2.5 miles) per hour. The story goes that two years later, another test ran out of control and the machine demolished part of the wall of the former royal arsenal building, which was by then the courthouse in Paris. Trials were abandoned, but Cugnot was given a pension by the King. Sadly that was taken back after the French Revolution, and Cugnot spent 20 years in exile, returning to Paris at the behest of Napoleon only shortly before his death. His was probably the first ever motorized vehicle, but few today even recognize his name.

PATENT POWER

Watt did not have the capital to build full-size machines. He went into partnership with the owner of an ironworks, but struggled to build a successful working steam engine. More than anything, he could not get a cylinder cast precisely enough to avoid the steam leaking badly. But he did apply for and was granted a patent for "a new method of lessening the consumption of steam and fuel in fire engines", with no more specific detail than that. This was such a loose definition of his invention that it meant he could claim ownership of almost any improvement to the steam engine that anyone else came up with; over the next 30 years Watt's life was dominated by rows over patent rights. His partner went bankrupt after the collapse of the Scottish banking system in 1772, but by then he had met Matthew Boulton, and the Birmingham entrepreneur took over the share of the business. After the death of his wife, Watt moved to join Boulton's Soho works in 1774 and a new partnership began. The problem of the leaking cylinders was solved through another Lunar Society connection. The armaments industry was burgeoning around this time, as a result of the growing need of the British army to maintain control of its colonial territories, and the ironmaster, "Iron Mad" John Wilkinson, brother-in-law of Joseph Priestley, had perfected a new cannon-boring machine. He adapted his machine to create a cylinder for the Watt engine. The result was an outstanding success, accurate "to the thickness of an old shilling", and the first working engines were set up in 1776 – one to power the blast furnace at Wilkinson's ironworks, the other to pump water at a colliery in Tipton.

> *"Boulton and Watt's engines used about 75 per cent less coal than did the old Newcomen engines."*

Below: 18th-century miners had to rely on human and animal power alone. The introduction of steam-powered pumps did a great deal to improve mine safety.

Boulton and Watt's engines used about 75 per cent less coal than did the old Newcomen engines, and the two men made their money by licensing the rights to erect and run the machines, charging royalties of a third of the savings made by the client. (Initially, savings were measured in coal, but later, when steam engines were installed in breweries, savings were calculated based on how many more horses would have been needed to do the same work. (This is where the concept of "horsepower" comes from.) This is where the trouble began. In Cornwall, the cut-throat competition of the tin mining industry meant that mines had little option but to go for the best possible pumping engine, as this was almost the only area in which they could make cost savings. But the wide terms of Watt's patent bred great resentment among mine owners and

Newcomen engine erectors, and Watt spent many of the next years travelling to and fro across the country trying to enforce his patent – not the life he had intended. The Cornish mining community, in particular, were a fairly wild lot to deal with; one of Boulton and Watt's agents, tasked with delivering a writ demanding unpaid royalties, was famously dangled by his ankles over a mine shaft while being asked if he still wanted to deliver it. Not only were royalties unpaid but other engineers copied the Watt design and built their own pirated versions. Even ironmaster John Wilkinson was caught selling his own "Boulton and Watt" engines, prompting Boulton to bring the manufacture in-house to the Soho factory.

Birmingham dominated the world of steam until the end of the century. As Boulton once put it to the diarist James Boswell, "I sell here, sir, what all the world desires to have – power." Steam engines were installed in ironworks, breweries, factories, and mills, anywhere that the power of a piston could be transferred to rotary motion, to drive belts, pumps, saws, trip hammers, or bellows. By the end of the century, when their patent rights finally ran out, some 450 Boulton and Watt engines were in use.

The modifications that Watt carried out to his steam engines over the decades were numerous: the centrifugal governor to control speed, the throttle valve, the double acting engine, the compound engine – all of these made the steam engine more efficient and consistent in its operation, and ever more powerful. But all of them were the product of engineering trial and error, empirical testing, or even commercial competition. For example, someone else had patented a crankshaft, so Watt had to come up with another way to convert piston power into rotary motion; he invented "sun and planet" gearing. Theory was nowhere to be seen. To be fair to Watt, he had used the recently discovered concept of "latent heat" – the amount of heat a substance needs to take in or release when it changes state, such as when steam condenses to water – to inform his ideas about the separate condenser, but the understanding of the science behind the steam engine was lagging far behind. Heat, for example, was at that time thought to be a substance in itself. In a sense, the question "What is power?" was of little importance to those concerned with working out how it could be used to make money.

"The question 'What is power?' was of little importance to those concerned with working out how it could be used to make money."

Following page: The arrival of the steam locomotive in the early 19th century would transform society through high-speed travel.

AGE OF STEAM

The struggle against the patent royalties in Cornwall, however, meant that other ideas about power began to emerge. The Trevithicks – father and son, both called Richard – were powerful tin miners in Cornwall. It was they who had dangled Watt's agent over the mine shaft in the royalties dispute. They copied and pirated Boulton and Watt engines, but Richard junior was also an inventor in his own right. In an effort to free them of the obligation to pay royalties, he found a way to avoid using the separate condenser. Watt's engines filled the cylinder with low-pressure steam, relying on the atmosphere to drive the piston; Trevithick's, on the other hand, used high-pressure steam to drive the piston, with the exhaust steam being vented into the atmosphere instead of the separate condenser. High pressure meant that everything about the engine was smaller and consumed much less fuel, and by the late 1790s Trevithick engines were being installed in Cornish mines for pumping and winding. But what made Trevithick's name in history was that he quickly realized that his engine was small enough to be mounted on wheels, along with both its water and coal supplies, and to drive itself. On Christmas Eve of 1801 the world was introduced to the "Puffing Devil"; Trevithick's full-sized, self-propelled steam carriage transported several men up Camborne Hill. Steam locomotion had arrived.

Richard Trevithick
1771–1833

Below: A series of illustrations of Trevithick's 1801 "road locomotive", the Puffing Devil.

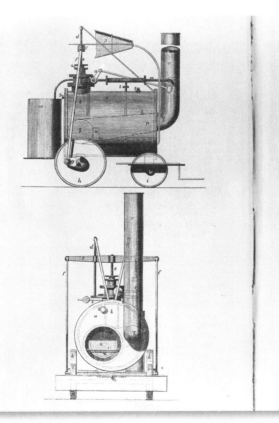

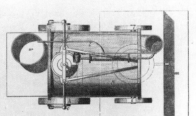

CAMBORNE COMMON ROAD LOCOMOTIVE.   127

TREVITHICK'S FIRST PASSENGER-CARRYING COMMON ROAD LOCOMOTIVE, CAMBORNE, 1801.

a, cylindrical boiler with wrought-iron ends, having inside it a wrought-iron tube bent as the l-tter U ; *p*, the fire-place, in one end of the tube ; *v*, fire-bars ; *s*, fire-bridge ; *z*, the ash-pit ; *g*, the return flue, leading to *r*, the chimney—the fire-door is not shown, as it would confuse the drawing ; *z*, the steam-gauge ; *s*, safety-valve ; *t*, soft metal safety-plug in top of fire-tube ; *j*, the bellows, blowing air into the close ash-pit, fixed to the guide-stays, and worked by the arm of its movable middle division connected with the piston-rod cross-head ; *b*, steam-cylinder let into the boiler, having a close top and bottom, with pipe for conveying steam to and from the bottom, and also the shell for the four-way steam-cock, and the steam-way from the boiler, all cast with the cylinder ; *o*, a four-way steam-cock, worked by a rod from the cross-head, with two tappets striking the lever, *o*, up and down, and having a handle, *o*, suitable for the engineman ; *k*, the feed pole-pump, worked from the cross-head ; *l*, the feed-pipe ; *w*, feed-water cistern ; *n*, case for heating feed-water by the passage of the waste steam through *w*, the waste-steam pipe, from the cylinder to the chimney ; *c*, the cross-head ; *f*, the two side rods ; *g*, the two cranks ; *h*, two driving wheels ; *i*, two steering wheels ; *e*, piston-rod ; *d*, guides for the piston-rod cross-head.

two front or steering wheels were turned by a rod conveniently placed close to the engineman attending at the fire-door.

One result of these experiments was the immediate application for a patent, granted on the 24th March, 1802, to Richard Trevithick and Andrew Vivian, for steam-engines for propelling carriages, &c., which may be read and studied by the young engineer with pleasure and profit even in "this" age of greatly-improved steam mechanism.

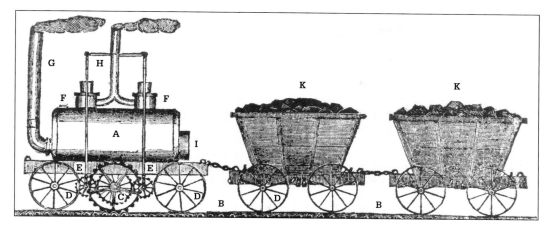

Above: A contemporary illustration of George Stephenson's 1814 locomotive, Blücher.

Description of the plate
A Boiler
BB Rail road
C The propelling wheel, which is put in motion by the agency of steam, or any first mover.
DD The carriage wheels

EE Connecting rods
FF Steam cylinders
G Smokey chimney
H Steam or discharging pipe
I Fire place
KK Coal waggons, or carriages of any

In 1804, a Trevithick engine was mounted on wheels and used to haul 10 tonnes of iron, five wagons, and 70 men over a 16km (10 mile) long iron "tramway" at Merthyr Tydfil in Wales. Trevithick went on to build a steam locomotive that he called "Catch Me Who Can", and even put on a steam circus ride in London, which was greeted with an indifferent response. But others took up the idea, and within a few years steam engines were being used to haul coal along tracks from collieries. Critically, the Watt patent expired at the start of the 19th century, allowing many others to try their hand at the design, manufacture, and exploitation of steam power, free from the worry of a Boulton and Watt lawsuit.

"For Britain, and the world, the Age of Steam had truly arrived, and transformed every aspect of industrial life."

For Britain, and the world, the Age of Steam had truly arrived, and the hissing power of high-pressure steam transformed every aspect of industrial life. Factories, fields, mills, and mines all used steam power to drive the rhythm of the working day. Factories were no longer tied to water or windmills and steam-powered trains provided new ways of transporting goods, and people. The steam engine also created a whole new language: "to blow off steam"; "a full head of steam"; "under your own steam"; "to run out of steam"; "to let off steam" – all of these were now used as often to describe human behaviour as the operation of machines.

Across mainland Europe, Britain, and America, energy was being released from coal on an unprecedented scale. Everything depended on more, faster, and more efficient machines, and it was in this world in the first half of the 19th century that theory at last began to catch up with the practical engineers who had learned how to tap the

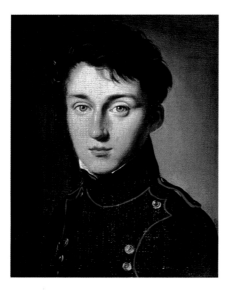

power of nature. Scientists arrived at what became known as the first law of thermodynamics. The first steps towards this coherent theory of energy emerged in France, with the writings of Nicolas Léonard Sadi Carnot, a military engineer who decided to study the steam engine, asking that age old question "Is the work available from a heat source potentially unlimited", or in other words, "Can we have unlimited power?" In a scientific environment that still regarded heat as an invisible fluid that flowed from one object to another when out of balance, his book of 1824, *Reflections on the Motive Force of Fire*, set out the critical law that heat always flows from a hotter to a colder object, and a measurement for the "work" done when it does so. He also designed an idealized engine to act as a theoretical model for the study of energy; that engine was a forerunner of the design of today's internal combustion engine. Carnot's ideas were developed further in Germany by Rudolf Clausius and in Britain by the great physicist William Thomson, later Lord Kelvin, and by the 1850s the concept of the "First Law" had been set out. It stated that energy can be neither created nor destroyed; that it can be converted from one form to another (so heat becomes work) but the same amount will always exist.

Left: Sadi Carnot 1796–1832. Son of a Napoleonic general, Carnot abandoned his military career for physics and made huge advances in the study of energy.

SATAN'S ENGINES

Richard Trevithick's high-pressure steam engine was a radical departure from James Watt's earlier design, or indeed any of the earlier steam engines. It was much more efficient in its use of heat, with the furnace and the piston all encased within the steam boiler itself. It avoided the need for a separate condenser, and it also meant that the piston could be driven by the injection of steam in both directions of its movement, doubling the power. But the design was risky, to say the least, because what it also needed was a boiler that could handle steam at something like 50 times atmospheric pressure without exploding. Indeed, many people thought that Trevithick was getting in league with Satan. Steam cannot be seen at high pressure – only when it condenses – so sudden explosions of experimental boilers for no apparent reason began to be talked of as devilish phenomena.

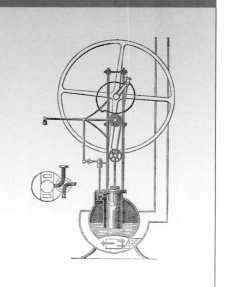

VICTORIAN EFFICIENCY

On its own, this was taken by some to mean that if only one could achieve perfect efficiency, with no loss of energy in friction, or heat, then perhaps, after all, a perpetual motion machine could be invented. Getting the most out of what you had was one of the driving principles of 19th-century society. Efficiency and economy became elevated to the position of higher moral values – a perfect fit with the Victorian mindset of thrift. At a domestic and everyday level, this was seen in the popularity of Mrs Beeton's famous *Book of Household Management*, which offered economical advice, recipes and tips on fashion, or how to manage your servants. Politically, the extent to which waste was seen as abhorrent is bizarrely illustrated by the ideas of James Thomson, brother of Lord Kelvin, who wrote bewailing the appalling cost of shipping every year from Peru some 200,000 tonnes of guano – bird manure – to be used as fertilizer and in gunpowder. Thomson pointed out that every day over 100,000 tonnes of human waste was tipped into the River Thames alone. Why not, he proposed, collect all that urine, boil it down and save our economy the wasteful cost of imports.

> **"Getting the most out of what you had was one of the driving principles of 19th-century society."**

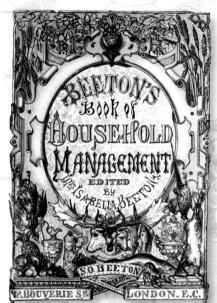

Above: First published in 1861, *Mrs Beeton's Book of Household Management* was far more than a cookbook – it set out principles for the efficient running of any household.

Virtue and frugality were intertwined in the Victorian mindset, so if only an engineer could be frugal and thrifty enough, then surely they could achieve limitless power from nature. The reality is that the concept of the conservation of energy put paid to perpetual motion, as no such loss-free device could possibly be attained in practice. But Victorian engineers could hope – in the second half of the 19th century over 500 patents for perpetual motion were lodged in the British Patent Office, compared with fewer than 25 in all the years before. Unfortunately for these hopefuls, the first law of thermodynamics was hotly pursued by the second, which states very simply that entropy will always increase. Entropy can be regarded as a measure of the amount of disorder in a system: be it a steam engine cylinder or the Universe itself. In practical terms, then, the second law means that heat always flows from hot to cold, as Carnot had identified, but more profoundly that everything in the Universe will gradually become more disordered. Energy will dissipate. Everything in the end must run down. That certainly put the proverbial nail in the coffin for perpetual motion. Some Victorian scientists hated it as a physical law, as it implied a hopeless moral decay, that no amount of effort could triumph in the struggle to pull order out of chaos; but what it did was to seed the notion that the Universe itself could have a beginning and an end.

SHOCKING FISH

Steam and heat had driven the industrial transformation of Europe, but just as power was beginning to be understood, another form of energy was making its first practical appearance in the world, and it leapt quite by surprise from the world of 18th-century parlour displays that had amused the members of Lunar Society and their friends. Colonel John Walsh, formerly secretary to the Governor of Bengal, Robert Clive (Clive of India), had amassed a fortune from his position in the East India Company, and settled back in Britain to become a Member of Parliament and to enjoy his main interest – science. He became fascinated by reports of the torpedo fish and, in 1772, went on an expedition to La Rochelle in France to capture and study the fish. There are 69 known species of torpedo fish; they are rays, and share an ability to deliver powerful electric shocks as and when they feel like it, either to stun their prey or to defend themselves. It is thought the fish were used by the Ancient Greeks to help with the management of pain, in childbirth or surgery. Walsh likened their ability to that of Pieter van Musschenbroek's early capacitor, the Leyden Jar, and even tried to build an artificial torpedo fish out of leather. The puzzle of the torpedo fish played into the debate over the ever fascinating phenomenon of electricity. Possibly inspired by Walsh's work, an Italian physician from Bologna, Luigi Galvani, had observed the twitching of frogs legs when accidentally struck with static electrical

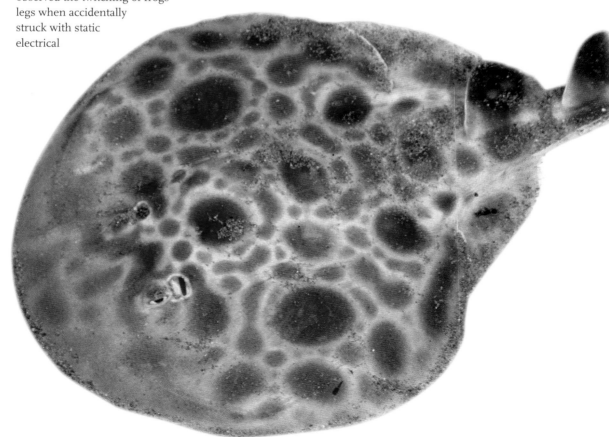

discharge, and had carried out a gruesome series of experiments to investigate what he called "animal electricity", convinced that it came from within the muscle of the animal. In this thinking he was opposed by Alessandro Volta, discoverer of potassium (see Chapter Two), who argued that the electricity was "metallic" and was outside the animal. Theirs was a long running competitive dispute of the type that still can happen in science today, but the net result was that Volta showed that he could also make the frog twitch by placing two different metals on either side of it, and closing the circuit with a wire. This was the beginning of the series of experiments that resulted in Volta's "pile", a little stack of metal discs,

> *"Magnetism and electricity, two mysterious phenomena that had fascinated philosphers for generations, were now very clearly linked."*

alternating between zinc and silver, separated by discs of paper soaked in brine. A wire at one end and a wire at the other, if touched together – on Volta's own tongue at first – produced an electric current. Volta realized the significance of what he had achieved and in 1800 wrote directly to the Royal Society in London, the most influential people he knew, to announce it to the world. His letter included a precise set of instructions for building a pile. This was the first practical battery. It had about the same power as an AA battery of today, and although it could last only a fraction of the time – until the paper discs dried out – was a truly mobile source of electricity. There was no longer any need to catch a lightning bolt, or spend hours rubbing glass tubes; Volta's battery produced not a spark that discharged in an instant, but a "current" that was available on tap.

Above: An individual "voltaic cell" consisted of a disc of copper and a disc of zinc, separated by a piece of cloth or cardboard soaked in brine.

Still no one knew what electricity was, but the scientific experimenters now took the lead. The Voltaic pile was refined to provide a bigger, steadier, constant supply of electricity to experiment with, enabling a range of new chemical discoveries, such as those that Humphry Davy achieved through electrolysis. Then, on 21 April 1820, one of those rather special observations in the history of science was recorded in the laboratory of a chemist in Denmark. The story goes that Hans Christian Oersted was preparing to give a lecture, when he noticed in his electrical apparatus that a compass needle had been deflected by an electric current passing through a wire nearby. The current appeared to create a magnetic field that ran at right angles to the wire. Magnetism and electricity, two mysterious phenomena that had fascinated natural philosophers for generations, were now very clearly linked – the concept of electromagnetism had arrived.

Left: Torpedo fish generate electric currents from "batteries" on either side of their head – their unusual properties had been known about since Roman times, when their shocks were used for treating pain.

INFORMATION FLOW

The world of electricity and the world of power were soon to become linked – although not in quite the way one might have expected. Oersted's finding caused a scramble among people with ideas to exploit the phenomenon, and what quickly emerged were several schemes to use the effect to create signals at a distance. By switching a current on and off it was possible to cause a needle to flick at the other end of the wire. Within little over a decade, several proposals for electromagnetic telegraph systems appeared in Germany, Britain, and America. Pavel Schilling, a Russian, demonstrated a telegraph of eight wires, with magnetic needles suspended on silk threads at the end, running between the rooms of his house; David Alter in Pennsylvania linked his house and his barn; and in Germany Carl Friedrich Gauss ran 1,000m (1,100yd) of wire over the rooftops of Göttingen. In Britain, a commercial telegraph was installed in 1839 along a section of the Great Western Railway, and six years later it enabled the police in London to be waiting for a runaway murderer who had boarded a train in Slough, some 40km (25 miles) away, when he arrived at Paddington Station. Here was a technology with an obvious practical benefit.

Samuel Morse
1791–1872

Perhaps 200km (125 miles) of railway lines could be travelled in Britain in 1830. By 1860, there were over 10,000km (6,200 miles). With the long straight tracks to run alongside, the telegraph wires rapidly transformed communication everywhere, but nowhere more so than in America. There, Samuel Morse's patented code, developed with his assistant Alfred Vail, became universally adopted, and as the great railroads carved their way across the continent, so the telegraph went with them. Thereafter, electricity has always been at the heart of information flow. The telephone, which converts sound to electric current and back, swiftly followed. Then came radio, which depends on the transmission of electromagnetic waves. And today's digital communication depends on the changing state of electrons.

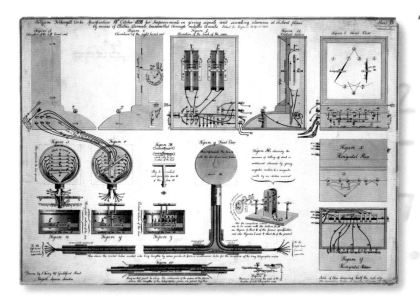

"The steam engine and the telegraph changed the world. Together they brought us the concept of universal time."

Left: Illustration from an 1838 patent for William Fothergill Cooke and Charles Wheatstone's first electric telegraph system.

The steam engine and the telegraph changed the world. Together they brought us the concept of universal time. Before it was possible to travel at the speed of steam there was no need for the time to be exactly the same in, say, London in the East of England and Cornwall in the South West. The world operated on a matter of hours, and it did not matter that Penzance, at the tip of the Cornish peninsula, ran eight minutes behind London. But now, precise time did matter. An eight-minute difference in the watches of two train drivers could easily result in a crash. The telegraph was quickly adopted by the railways; it was the only way to get a message to a station, or a signalman, before the train itself arrived. To ensure the whole network would run smoothly, all the trains began to adopt railway time – eventually referred to as Greenwich Mean Time. By 1860, almost every public clock in Britain showed the same time, and over the next few decades other countries around the world followed suit. In the USA, thousands of local "noons" gave way first to railroad times set by the headquarters of the rail companies and then to the standard time zones we know today.

With the telegraph, runaway criminals could be caught and runaway brides could be stopped before they took their vows; information moved at astonishing speeds not only across nations but across empires. By the 1870s, every continent was linked by cable, and keys tapped code from one side of the world to the other. The nature of the empires themselves was changed. Garrisons were needed to guard and secure the lines of cables and plantations of gutta-percha trees were grown to maintain the supply of their sap that formed the latex used in insulating the thousands of miles of electrical wire now threading around the globe. The fortunes of countries like Malaya, where the trees were grown, were transformed in consequence, as their value to the British economy changed. And above all, ever greater quantities of coal were needed to fuel the thousands of steam engines that now powered industry and transport.

THE TELEPHONE

The success of the telegraph meant that communication of sound at a distance became a clear goal. The invention of the telephone is usually credited to Alexander Graham Bell, a Scot living in America, who obtained the first patent for the device in 1876 and went on to commercial success. However, it is argued that the Italian Antonio Meucci demonstrated an early telephone some years before. Meucci had developed an electric "treatment" for rheumatism, and while electrocuting one of his patients he heard a scream seemingly passing down the copper wire – the sound vibrated an electrical conductor near the patient and created an electrostatic charge that in turn vibrated the electrical conductor near Meucci's ear. Meucci went on to develop a device with electromagnets linked to diaphragms. The principle was that sound moved the diaphragm, which moved the electromagnet, which in turn fluctuated the current down a wire. At the other end, the process in reverse recreated the sound. No working models of Meucci's device survived, and he failed to renew the payment for his patent application. It remains disputed as to whether or not he was the first inventor of the telephone.

EXPERIMENT AND PRACTICE

It was not long after Oersted's demonstration that electricity could move a magnet that the opposite relationship was seized upon. In London, Michael Faraday had been an apprenticed bookbinder, but had tried every possible approach to get a job working in some capacity in science, about which he was passionate. Chance events at the Royal Institution, including Humphry Davy temporarily blinding himself in a chemistry experiment, and the laboratory assistant being fired for assaulting the instrument maker, resulted in Faraday becoming Davy's new laboratory assistant. Faraday rapidly became Davy's key experimenter in electromagnetism and in 1821 managed to build a device that enabled an electric current to produce the circular motion of a magnet. This was the start of work that eventually led him to produce an electric motor. More significantly, ten years later, Faraday designed a machine that did the reverse: a moving magnet created an electric current in a coil of wire that was wound around it. This was the principle behind the dynamo – a constantly turning motor, such as a steam engine, could generate an electric current. What Faraday had done was to create both halves of the electrical power industry: electric current can generate movement; and movement can generate electricity.

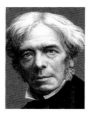

Michael Faraday
1791–1867

Dynamos and electric motors at first remained pretty much within the confines of the scientific laboratory or were used as curiosities to impress the public. But by the 1870s, a dynamo had been built that could supply electricity on an industrial scale, this time by a Belgian engineer, Zénobe Théophile Gramme, who also built a near identical machine operating on the reverse principle to act as a motor. With industrial-scale electricity now available, factories could use steam or water power to turn huge dynamos and generate electrical power, which they could then distribute to smaller electric motors to drive their machines, rather than the complex belts and drives that were needed to connect directly to steam engines themselves. By the end of that decade, the incandescent light bulb had also been successfully produced to a commercially viable standard by Thomas Edison, so that the workplace could now easily be lit well into the night. The new electrical technology ushered in what has become known as the "Second Industrial Revolution". Countries such as Germany, who had come late to industrialization, were now able to develop their economies rapidly, and the industrial and commercial power balance in Europe began to shift away from the dominance of Britain.

Right: An electrified factory at Hanover, Germany, in the early 20th century. The introduction of electric power was an important element of the so-called "Second Industrial Revolution".

Right: Electric motors use the repulsion and attraction between fixed magnets and the changing electromagnetic field in a coil of wire to turn a rotor and spin a drive shaft.

PERSONAL POWER

Yet by the end of the 19th century, it was still unthinkable that electricity would come to dominate the planet as it does today. For all the changes at the factory, walking home at night still meant walking through a world illuminated via the burning of coal gas. Electricity offered huge potential to change all that, but it suffered from one major limitation: distribution. It was all very well to have a steam engine next to a factory to turn the generator, to turn the motors, and light the light bulbs. But the energy lost along the wires was so great that it would require a steam engine and generator at the end of every street to service a town. Indeed, early distribution networks had a limited run of about 2km (1 mile). The challenge was to come up with a system that could offer useful electricity at the end of a very long wire.

In 1883, the Free Niagara Movement was triumphant in its campaign – one of the first environmental movements – to return Niagara Falls to a more natural state than was then experienced by every visitor to the area. The vast natural power source of the waterfall that lies on the Canada–USA border had been tapped since the first settlers cut mill races into the banks of the river a century or more before. But by now the shores on both sides were cluttered with mills and factories of every sort, all driven by the relentless flow of the water. The creation of a state reserve around the falls meant that the commercial free-for-all was brought to an end, but the bulk of the power of the falls remained untapped – clearly a terrible waste of commercial opportunity. Then, in 1886, an engineer on the nearby Erie Canal, one Thomas Evershed, proposed a huge engineering project to build a series of tunnels and channels to carry the power of the waterfall away from the reserve, so that it could be used. The potential cost was astronomical, and yet it was not obvious how the power could be exploited in order to get a return on such a huge investment. The Falls generate 8,000,000 horsepower, far too much for the needs of the nearby small town of Niagara Falls itself, which had a population of just 5,000, so the question was asked: could it somehow be distributed to the growing town of Buffalo (population 250,000 and rising) some 40km (25 miles) away?

Right: Up to 5,700 cubic metres of water pass through the kilometre-wide falls at Niagara every second, making them an irresistible source of power.

POWER STRUGGLE

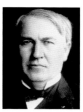

Thomas Edison
1847–1931

Two giants of the electricity industry provided two different answers. Thomas Edison, inventor of the first practical light bulb and many other electrical devices, had set up small local electrical distribution networks in Manhattan, using direct current. This is a form of electricity where the current flows continually in one direction, and it is subject to great loss due to the resistance of the wire along which it travels. To travel more than about 2km (1 mile) either the current at the start has to be so high as to burn out any light bulbs or motors, or the cables have to be too thick to be practicable. On the other side of the competition, George Westinghouse, inventor of the compressed air braking system for railways, had bought up patents belonging to Nikola Tesla, a disaffected former employee of Edison, who had come up with key improvements to the concept of alternating current – where the electric current literally flows in alternate directions along the wire. The idea behind this is that higher voltage is used to transfer the current for great distances, while a transformer – for which Tesla's inventions were critical – steps the voltage back down to a more manageable level at the consumer end.

The battle between these two cut-throat entrepreneurs was the high point of the "War of Currents" that had been running for over a decade, with different camps across Europe and the USA claiming their system was best. Edison tried every marketing trick in the book to discredit alternating current, including emphasizing its high voltage dangers, by deliberately promoting the use of Westinghouse's system for the newly invented electric chair. In the end, the first successful long-distance distribution of electricity was demonstrated in 1891 at the International Electro-Technical Exhibition in Frankfurt in Germany. The electricity was generated as alternating current at a cement works in Lauffen am Neckar, 175km (109 miles) to the south, and carried on overhead wires to light a display of a thousand bulbs at the entrance to the exhibition and, ironically, to power the turning of an artificial waterwheel. The spectacular success of this event was a deciding factor in the plan for Niagara, and it was alternating current that won the contract. At midnight on 16 November 1896, Tesla and Westinghouse's system proved itself; power from their transformers reached Buffalo. The first 1,000 horsepower went to the street railway company, and the local power company had immediate orders from residents for 5,000 more. Within a few years the number of generators at Niagara Falls grew to ten, and power lines were electrifying New York City. Broadway was ablaze with lights; the elevated street railways and subway system rumbled. Electrical power, the crucial element for nearly all modern technology, would soon be everywhere.

"Electrical power, the crucial element for nearly all modern technology, would soon be everywhere."

Left: Reliable electric lighting transformed cities in Europe and America, making the streets safer and helping to foster a 24-hour society.

Right: Edison's incandescent light bulb was not the first of its kind, but a number of innovations made it more robust and able to function within a power network.

RADIOACTIVE MAGIC

The search for new sources of power had changed the face of the modern world. Electrical energy was capable of nearly anything. But there was a catch. Unless you had a waterfall on hand, to generate the electricity you needed, the same old-fashioned sources of power – whether it was oil, coal, or even wood – needed to be burnt to power the generator. Then, right at the turn of the 20th century, a whole new source of energy emerged, which briefly rekindled the dream of unlimited power. Henri Becquerel had discovered it, and Marie and Pierre Curie had studied it: radioactivity. Radioactivity appeared to produce heat and light from nowhere. In the words of William Crookes, inventor of an early electrical discharge tube (see Chapter Two), this new power was "an example of seemingly continuous energy – something of which we had previously no conception – who can tell what fresh achievement it may be the forerunner?" Certainly Pierre Curie thought that it could continue forever.

The Curies discovered radium, a product of the radioactive decay of uranium, in 1898. The strange new energy took popular imagination by storm – for good and, as it turned out, also bad. Today, locked out of harm's way in a special room in Tennessee, there lies a collection of the many products based on its remarkable properties, including the "spinthariscope", invented by Crookes. It contains a tiny piece of radium, too small to be seen by the naked eye, and a miniature fluorescent screen at which you were invited to look, through a magnifying brass eyepiece, in a dark room, to see an "endless display of shooting stars". There are also radioactive toothpastes, beauty creams, pills, and radium tonics, called liquid sunshine, which were said to keep the drinker healthy. Indeed animals and people did seem to thrive after a short exposure to radium, but it turns out that this was only a side effect of the body over-producing red blood cells, a natural defence mechanism against the destruction of the radiation poisoning. One Pittsburgh industrialist drank a brand of radium water called "Radithor" every day, even sending crates of it to his friends. He died painfully, with the bones in his jaw decaying.

"We just don't know. We have made a discovery of forces and power beyond present knowledge, quite beyond imagination."

Below: Newly discovered radioactive materials were used for many patent medicines and similarly untrustworthy products, causing untold harm before the dangers were realized.

CRÈME
SCIENTIFIQUE

CURATIVE
EMBELLISSANTE

THO RADIA

THO - RADIA
à base de thorium et de radium selon la formule du
DOCTEUR ALFRED CURIE
EN VENTE EXCLUSIVEMENT CHEZ LES PHARMACIENS

Above: Even as late as the 1930s, cosmetic products using radium were still being marketed.

The simple fact was that no one, not even its discoverers, had any idea how it worked. When Pierre Curie was asked what had been discovered he replied: "We just don't know. We have made a discovery of forces and power beyond present knowledge, quite beyond imagination." Yet within just a few years it would be recognized as a display of the fundamental principles of energy at work.

RADIOACTIVITY

The chemical elements that make up all matter are arranged in the Periodic Table in order of their atomic number – the number of protons in the nucleus. The lighter elements are stable, but the heavier elements are unstable. In a heavy element, such as uranium, the force binding neutrons to protons in the nucleus of the atom is insufficiently strong to hold them together, and so the nucleus undergoes radioactive decay, emitting subatomic particles and electromagnetic radiation: so called alpha, beta, and gamma radiation. The decay continues, with mass being lost until the element is transformed into a stable form. Uranium (atomic number 92) eventually decays to a stable form of lead (atomic number 82). The time taken for half of the nuclei of an element to decay is known as its "half life". The half life of the most commonly occurring isotope of uranium is about 4.5 billion years. So although the process will emit energy for a very long time, it will eventually come to a stop, and thus does not violate the laws of thermodynamics. The name "radioactivity" was coined by Marie Curie.

THEORY CATCHES UP

Theory did in the end catch up with the practical exploitation of power. Michael Faraday, as he studied the effects of electricity and magnetism, came to believe that there was a hierarchy of forces, with electricity – God's force – at the top, and gravity somewhere below. Faraday was the supreme experimenter of his age, and showed conclusively that electric current, electrostatic charge, and magnetism were at heart the same phenomenon. He argued that electricity and magnetism acted along "lines of force", taking time to move, perhaps in the form of wave motion, but he did not have sufficiently good mathematics to take his ideas of magnetic and electrical "fields" any further. Instead, it was a Scottish mathematician, nicknamed "Daftie" at school but regarded by many as the greatest classical physicist since Newton, who bound the ideas together as mathematically precise laws. James Clerk Maxwell was brought up in Edinburgh, attended university there and at Cambridge, and became a professor at Aberdeen at the astonishingly young age of 25, before settling at King's College in London. He worked on the physics of colour and the nature of the rings of Saturn, but his crucial work was to take Faraday's ideas of force fields and show mathematically that electricity, magnetism, and also light itself were manifestations of exactly the same phenomenon – electromagnetic waves. In 1888, nine years after Maxwell's death, this was proved conclusively when the German physicist Heinrich Hertz demonstrated the existence of radio waves, and showed that they travelled at the speed of light.

James Clerk Maxwell
1831–1879

Through the notion that the speed of light is always constant, Maxwell's equations also provided the basis for the radical ideas that emerged from the mind of Albert Einstein as his theory of special relativity. Published in 1905, it turned physics on its head, running counter to the existing Newtonian mechanical vision of the Universe and introducing counterintuitive concepts such as space-time, length contraction, and time dilation into the physicist's vocabulary. But it also provided a fundamental explanation of what power actually is, in the most famous equation in science: $E=mc^2$. Energy equals mass, multiplied by the square of the speed of light, an almost unimaginably large number. Until this point energy had been understood as heat, electricity, even

Below: The nuclear fusion reactions at the heart of a hydrogen bomb explosion unleash incredible destructive force – and properly harnessed, they may also offer a solution to our future energy needs.

wind or water power, and the different forms could be transformed into one another. But this was a far stranger level of equality; heat, motion, and radiation could all now be seen as expressions of this fundamental insight. Mass – the material of anything – contains almost incomprehensibly vast amounts of energy, which can be released by motion, burning, compressing, or splitting.

None of the great inventors or engineers of the previous three centuries had needed that equation to build the machines that built empires or transformed nations, but now that it was there it explained all their work. It explained too the awesome power that could be released by the splitting of the atom. Radioactivity suddenly could be seen as the raw energy released by infinitesimally small amounts of mass simply decaying. So for the first time, scientific theory set a goal for the inventors and engineers to aim for. Einstein himself, at first, was certain that such power could never practically be achieved, but of course we know now that in this he was wrong. The atomic bomb was created, and today's nuclear arsenals are testimony to the achievements of the science.

"Whatever form of energy we tap, it almost always comes to us via the Victorian technologies for generating and distributing electricity."

There is more irony in the story of power. Today, nuclear radiation is a major source of energy for generating electricity, but the radioactivity in nuclear power stations is simply another fuel like coal, gas, or oil. It creates heat, to make steam, to drive turbines, which turn generators. This is true for most of the latest means of providing energy, be it renewable sources like wind or waves, hydroelectric, or even green fuels such as biomass. Whatever form of energy we tap, it almost always comes to us via the Victorian technologies for generating and distributing electricity.

So it was the search for limitless power that gave us the practical machines that drive the modern world. At the same time, the attempts by experimenters and theoreticians to find out how those machines worked revealed deep truths about science. They arrived at equations that reveal the vast potential of energy that exists in the world, and the laws of thermodynamics that determine why none of it can in fact be limitless.

CLASSICAL GREECE

ROMAN EMPIRE

MIDDLE AGES

ISLAMIC SCIENCE

AGE OF DISCOVERY

RENAISSANCE

REFORMATION

SITVS
IRCVLIS
CIRCVN

^ First battery

^ Leyden Jar

The story of creating power begins with the exploitation of the most obvious natural resources around us – water and wind. In the Age of Enlightenment, two other sources emerged. Electricity was a fleeting charge created by friction until in the mid 1740s Pieter van Musschenbroek invented a way of storing it, a capacitor or "Leyden jar". Alessandro Volta improved this still further with his "voltaic cell", the first practical battery.

At the same time the age of steam began. Primitive steam engines had existed for centuries, but Thomas

Alessandro Volta
1745 – 1827

James Watt
1736 – 1819

James Clarke
Maxwell
1831 – 1879

Michael Faraday
1791 – 1867

Thomas Edison
1847 – 1931

EARLY 20TH CENTURY

MID 20TH CENTURY

AGE OF ENLIGHTENMENT

21ST CENTURY

^ Early electric motor

^ Lightbulb

^ Telephone

^ Atomic explosion

Two related inventions of Michael Faraday made electricity into a power source that would harness and ultimately eclipse steam – the electric motor and, more significantly, the electric generator. By the late 19th century, the "Second Industrial Revolution" began, with innovations such as the lightbulb (first reliably built by Thomas Edison) and the telephone changing the way we lived, worked and communicated for ever.

The last act of the story is where the overwhelmingly practical nature of the search for power connects with the theory, James Clerk Maxwell's research into electromagnetic waves paving the way for the work of Albert Einstein and the awesome power of the atom.

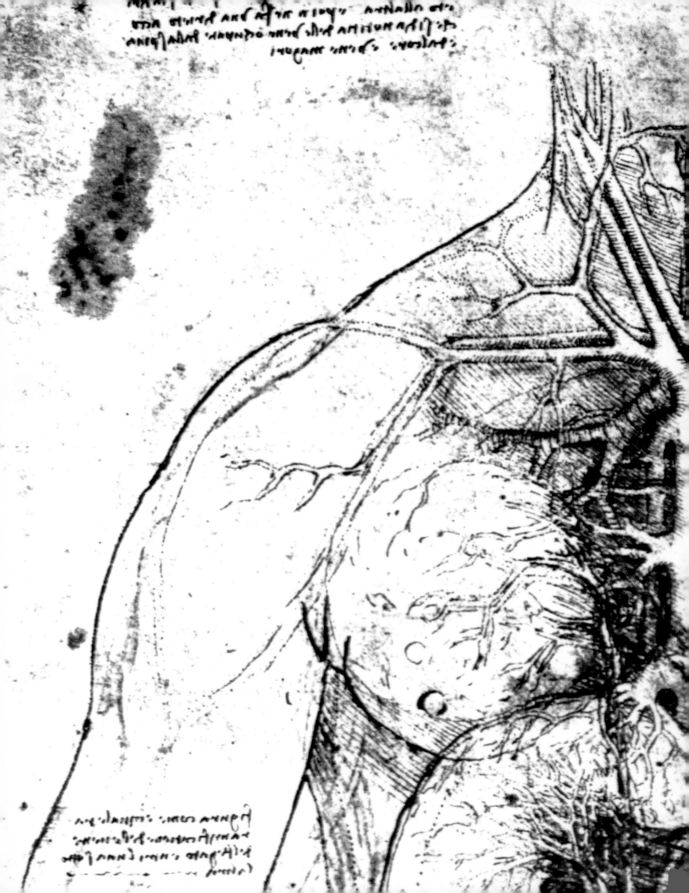

Body

WHAT IS THE SECRET OF LIFE?

Go out shopping and buy 18kg (40lb) of carbon, enough phosphorus to make two thousand matches, and a small iron nail. Then drop in on a friendly chemist and collect small amounts of a few other relatively common elements. Take it home, mix it in a bucket, add about 50 litres (11 gallons) of water, and stir. The resulting mixture is chemically similar to a person. Yet it is not a human and cannot, of course, be brought to life.

So what is the secret to life? What is it that turns a pile of chemicals into a living, breathing biological entity? The hunt for answers to these questions has created modern medicine and allowed us to reach a point where it seems that life itself will shortly be synthesized in the lab. The moment we create an entirely manmade cell is not that far away and when it happens it will be one of the most extraordinary in our history.

Over the last couple of centuries in the West, two rather different approaches have been employed in trying to find out exactly what it is that makes us tick. The first has been simply to cut things open and take a look inside. Is life, perhaps, simply a product of the way we are put together? This approach has been productive, but we may now have reached the limits of what further dissection can tell us. The second approach has been to look for a life force – something physiological that could explain the difference between a dead body and a living one. This approach, as we will see, led to important discoveries like biological electricity and the role of hormones, but in itself it did not bring us any closer to answering the question "What is life?"

Left: A detail from one of Leonardo Da Vinci's anatomical sketches of a human torso. Despite such detailed anatomical studies, figuring out the function of the various organs would be a long struggle.

THE ANATOMISTS

The first person to create accurate drawings of the human body was the Renaissance genius, Leonardo da Vinci. The illegitimate son of a lawyer and a peasant woman, Da Vinci was not given a formal education; perhaps because of this he was more questioning, more willing to rely on his own eyes than on the wisdom of the ancients. When he drew, he drew what he saw, not what others told him he should see. His paintings relied on new studies in perspective, but they also relied on a complete understanding of the human body. When he was first apprenticed, his master, the painter Andrea del Verrocchio, insisted that he and all his fellow students study human anatomy. Later, when he was a famous artist, he gained more intimate insights into the human body by dissecting corpses. At the time – the early 16th century – this was not illegal, but it was frowned upon and undoubtedly grisly. He dissected mainly at night, with the help of a young assistant, and was often in a race against time and the rapid onset of decay.

Leonardo Da Vinci
1452–1519

Da Vinci's anatomical drawings are amongst the finest ever produced and include the first drawing of a foetus in utero. But what is particularly striking about Da Vinci's work is that he not only drew what he saw but also reached insightful conclusions from what he observed. By comparing the arteries of an old man with those of a young boy, for example, he concluded that the furring up of the old man's arteries had contributed to his death. He had effectively described atherosclerosis hundreds of years before anyone else and at a time when many doctors thought that arteries carried nothing but air. Unfortunately, as with much of his work, he never finished it to his own satisfaction – there was always something else to find out about – and it was not published for another 160 years.

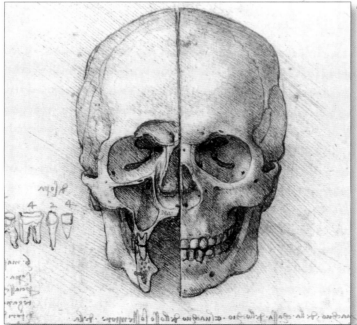

"Da Vinci's anatomical drawings are amongst the finest ever produced and include the first drawing of a foetus in utero."

Left: Around 1489, Leonardo completed a series of detailed drawings of the skull, attempting to locate the seat of specific mental faculties within the brain.

Instead, in Da Vinci's day, doctors relied for their understanding of human anatomy on crude drawings from the 2nd century by a man called Claudius Galenus, also known as Galen. A Greek, born sometime around AD 130 in what is now Turkey, Galen had begun his career treating gladiators, which must have given him an intimate knowledge of the human body in extremis. Because dissecting human bodies was actively discouraged at that time, his studies were confined to animals, from monkeys to pigs, and he took great pleasure in cutting open baboons in front of interested Roman citizens to show off his great anatomical knowledge. His textbooks were seen as almost sacred texts and never questioned. It was not till long after Leonardo's death that a Flemish dwarf called Andreas Vesalius would finally change that.

Right: Greek physician Galen was the ultimate authority on all medical matters for almost 1,500 years.

CHINESE MEDICINE

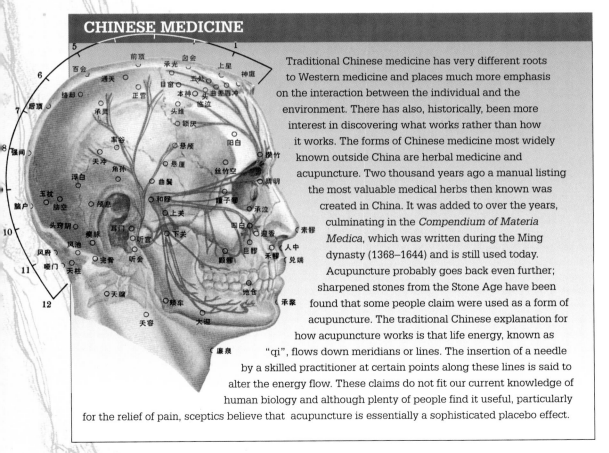

Traditional Chinese medicine has very different roots to Western medicine and places much more emphasis on the interaction between the individual and the environment. There has also, historically, been more interest in discovering what works rather than how it works. The forms of Chinese medicine most widely known outside China are herbal medicine and acupuncture. Two thousand years ago a manual listing the most valuable medical herbs then known was created in China. It was added to over the years, culminating in the *Compendium of Materia Medica*, which was written during the Ming dynasty (1368–1644) and is still used today. Acupuncture probably goes back even further; sharpened stones from the Stone Age have been found that some people claim were used as a form of acupuncture. The traditional Chinese explanation for how acupuncture works is that life energy, known as "qi", flows down meridians or lines. The insertion of a needle by a skilled practitioner at certain points along these lines is said to alter the energy flow. These claims do not fit our current knowledge of human biology and although plenty of people find it useful, particularly for the relief of pain, sceptics believe that acupuncture is essentially a sophisticated placebo effect.

DEATHLY DISSECTIONS

Picture the scene: it is 1536 and a criminal has been hung and left to rot on the gibbet. One evening, along comes Vesalius, a 22-year-old medical student. He looks up at the corpse with longing; human dissections may not be illegal, but body snatching most definitely is. That does not deter him. Vesalius jumps up, grabs the legs and pulls. With a terrible ripping sound they come off in his hands. He runs away into the night, clutching them in his arms. Later, he returns for the rest of the body.

What Vesalius was doing was extraordinarily dangerous. Not only was he risking jail and personal ruin, but his meticulous dissections of this and other bodies would challenge a belief system that had remained in place for centuries. At Vesalius's medical school there was just one set of anatomical textbooks – those written by Galen, whose words would be read out by the senior doctor while unquestioning students looked on, nodding, from a distance. Vesalius, however, had decided to do something that would have outraged and disgusted his contemporaries: to dissect and examine a human body himself.

Undeterred by the rotting human flesh, Vesalius put the stolen body on his kitchen table and set about stripping it down to its bare bones. He treated the task as if he was making beef stock. First, he filled a big pan with water and set it to boil. Then, he took bits of the corpse and removed as much skin and flesh as he could before dropping them into the pan, where they boiled away for hours, until falling apart. Finally, bone by bone, he tried painstakingly to identify every single part of the human skeleton. It was an arduous task; there are 206 bones in the human body, and this was just the start. Vesalius wanted to map not just the bones, but also every organ, ligament, and muscle, and was determined to understand where they all fitted – it was like putting a jigsaw back together. To do this he needed more bodies. So, naturally, he stole them.

Left: Vesalius's detailed anatomical drawings overturned many preconceptions that had remained unchallenged for over a millenium.

Below: Andreas Vesalius 1514–1564. Born into a family of physicians, the great anatomist spent much of his childhood dissecting birds, mice, and other small animals.

"Vesalius correctly identified the location of all the major organs, nerves and muscles in the human body and, significantly, began the proper study of human anatomy."

Above:
Illustrations from Vesalius's work, including the frontispiece illustration of a theatrical dissection (left).

Later, as chair of Surgery and Anatomy at the University of Padua, he found a sympathetic judge who was able to provide him with a supply of bodies obtained in more legitimate ways. Despite being a professor, Vesalius continued to carry out dissections himself and employed talented artists to illustrate his work. These drawings were eventually put together in 1543 in the form of a book, *De Humani Corporis Fabrica* (On the Fabric of the Human Body). In this book, he pointed out that much of what Galen had written was incorrect, and that 1,300 years of medical teaching were seriously flawed. Some dismissed him as a madman. Others, who took the time to read more carefully, wondered if perhaps the human body had changed since Galen's time. In all, Andreas Vesalius corrected over two hundred of Galen's mistakes.

Vesalius's masterpiece was published in the same year as Copernicus's great work, *De Revolutionibus* (see Chapter One). Thus, 1543 is regarded by some as the start of the scientific age, although this could be considered a little farfetched, since Copernicus's book would not have any significant impact on the world till long after his death. *De Fabrica* was not covering a subject that was quite as epic in its scale as the realignment of the Solar System, but it was in its own way a truly monumental piece of work. In it, Vesalius correctly identified the location of all the major organs, nerves, and muscles in the human body and, significantly, began the proper study of human anatomy. He had shown what he and others like Da Vinci passionately believed – that the wisdom of the ancients could not be relied upon and that experiment through observation was what was needed.

Sadly, Vesalius did not enjoy a long and happy life. He fell out with a patron and embarked on a pilgrimage to the Holy Land. (Some say his conscience got the better of him and he went to atone for all the bodies he had desecrated, though it is extremely unlikely that he ever had such qualms.) On his way home in 1564 he was shipwrecked on a Greek island where he starved to death. He was just 50 years old.

THE BODY AS MACHINE

The University of Padua, where Vesalius did much of his pioneering work, continued to build on its reputation as a leading and forward-thinking medical centre. In 1597, 54 years after *De Fabrica* was published, a young doctor from England called William Harvey went to study there. Harvey was the son of a farmer and had studied medicine at Cambridge, but found the teaching dull and uninspiring and so moved to Padua. There he started to develop ideas that were every bit as unsettling as those of Vesalius. The irony is that William Harvey was not looking to change the status quo – he was a traditionalist at heart. He took a long time to publish his findings, partly because he feared ridicule and partly because he worried about how they would be interpreted. Like Copernicus, he was a reluctant revolutionary, forced into a situation in which he could not help but destroy one of the few elements of Galen's work that remained unquestioned. In doing so, he started a movement that would firmly establish a mechanistic view of the body, something that as a God-fearing traditionalist he would have hated.

When Harvey arrived in Padua he studied under Girolamo Fabrizio, also known as Fabricius. Fabricius, a surgeon, is now chiefly famous for being the first to describe the valves inside veins. Although he created detailed drawings of the valves, he was completely mistaken about their purpose. He thought that the valves were there to control the rate at which blood flows from the liver to the other organs. In this he was clinging to yet another of Galen's mistakes. Harvey would later demonstrate the real role of the valves and, by showing that blood circulates through arteries and veins, fatally undermine the Galen world view.

Below: The University of Padua was a great centre for medical and scientific research between the 15th and 18th centuries.

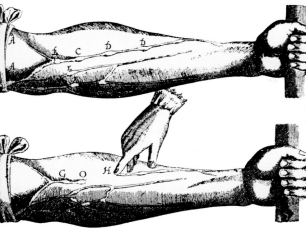

In Harvey's time, blood itself was seen as one of the four humours – the four basic substances that were thought to fill the human body. According to 16th-century medicine, these substances were phlegm, yellow bile, black bile, and blood. The critical thing was to have a balance of these humours; too much blood, for instance, could cause sickness. And if you believed that then bloodletting, an extremely popular medical practice as late as the 19th century, was a completely logical treatment. But the way that doctors viewed blood – what it is and what it does – was still largely dependent on ideas that were extremely ancient. It

Above:
Illustrations from Harvey's book demonstrate the direction of blood flow through the veins of a human arm.

was widely accepted, and consolidated by Galen's teachings, that humans have two different blood systems, arteries and veins, which are not connected and which have completely different roles in nourishing the body. According to Galen, the liver made blood, which travelled to the rest of the body via the veins and was totally consumed in the course of its journey. The arteries, on the other hand, carried something called "vital spirits" from the lungs to body, this being their primary task. This was clearly wrong, but not a bad guess considering that oxygen would not be discovered until Priestley and Lavoisier's experiments in the late 18th century (see Chapter Two).

ISLAMIC MEDICINE

In medieval times, the practice of medicine was vastly more sophisticated in the Islamic world than in the West. Islamic scholars collected and then translated into Arabic huge amounts of written work that had been passed down from both the Ancient Greeks and from the Indians. They not only made these writings more accessible and comprehensible, but also built on what they contained. One of the greatest physicians of the Islamic world was a man now known as Avicenna, who lived during the 11th century. Among other things, he realized that diseases like leprosy are contagious and that they can be spread by close contact, and emphasized the importance of quarantine for controlling the outbreaks of plague. His books were widely used in European universities well into the 17th century. Muslim doctors created what might be described as the first modern hospitals, where the sick were looked after by properly trained physicians. They were clean and organized; by contrast, hospitals in Christian Europe were disorganized, filthy, and provided few effective treatments beyond bed rest. Thanks to the work of Avicenna, Islamic hospitals were created with separate wards, so that patients who had contagious diseases could be kept apart from other patients.

THE EXPERIMENTALISTS

Returning to England in 1602, Harvey's society connections enabled him to become a member of an elite club, the Royal College of Physicians. In his new role as a Fellow of the College, he was expected to give a three-day lecture on physiology for its esteemed members. Since it was difficult – or rather, impossible – to demonstrate the physiological processes of living bodies in corpses, Harvey had to be inventive. He introduced animal vivisection into his lectures, and demonstrated what he safely could on humans. His search for suitable demonstrations led him to explore the relationship of the lungs and heart, which focused his attention on the heart and, in turn, on blood.

Harvey believed, like almost everyone at the time, that blood was created continuously in the liver. But how could he demonstrate the hepatic production of blood to the members of the Royal College? Slowly, he came to an awful realization: he could not, because everything he tried showed clearly that what Galen had claimed was utterly wrong. Under close scrutiny, Galen's description of the way that blood is created was impossible to accept. Harvey had started simply enough by measuring the capacity of the heart and working out how much blood it must be pumping per minute, but to his horror, even using the most conservative estimates, he had arrived at a figure of well over 240kg (530lb) of blood a day – more than three times the entire body weight of a man. This made no sense at all. He worked and reworked his calculations, but the numbers were always the same. He reluctantly concluded that Galen must be wrong and that the body could not possibly be making so much blood and destroying it in such a short time scale.

Further experiments soon convinced Harvey that there was a far more plausible explanation for what was going on inside the body. Arteries and veins must be connected and forming a circulating system. Unfortunately, microscopes powerful enough to pick out capillaries, the small vessels that link the arteries and veins, did not yet exist, so Harvey had to try and prove his theory by indirect methods. Some of his most famous sets of experiments were those he performed on himself. First, he tied ligatures around his upper arm until he had cut off all the blood flow to his fingers. Then he slowly loosened the ligatures until the veins in his lower arm (below the ligature) began to bulge. He explained that this was happening because when he loosened the ligatures, blood from the arteries could now reach his fingers. But this blood could not drain through the veins back to the heart because it was still being blocked by the ligature; veins, as we now know, are shallower than arteries and need

Below: The flow of blood between organs according to Galen, with blood moving through the body in tides rather than circulating.

Right: Illustration of an early attempt at a blood transfusion, replenishing a person's blood supply with that from a lamb.

Left: William Harvey 1578–1657. Harvey's new approach to medicine helped to establish a new paradigm of experiment and observation rather than reliance on old authorities.

EXERCITATIO. ANATOMICA DE MOTV CORDIS ET SAN- GVINIS IN ANIMALI- BVS, GVILIELMI HARVEI ANGLI, Medici Regii, & Professoris Anatomiæ in Col- legio Medicorum Londinensi.

FRANCOFVRTI, Sumptibus GVILIELMI FITZERI. ANNO M. DC. XXVIII.

Left: Frontispiece of Harvey's revolutionary book. While his findings were controversial, they did little to affect his reputation as a physician.

less pressure to block them. He also deduced the real purpose of the valves that his former tutor had discovered: they are there to ensure that blood flows – and particularly when it has to do so against the force of gravity – in one direction, back to the heart.

As a result of these and other experiments, Harvey was soon convinced that blood must circulate round the body, driven by the heart. Strong in his religious and Aristotelian beliefs, Harvey claimed, "The concept of a circuit of the blood does not destroy, but rather advances traditional medicine." He was completely wrong – his discovery utterly undermined traditional teaching. It would help launch a new way of seeing the body, not as a balance of vital forces but as a complex mechanical machine. Rightly nervous about how his findings would go down, however, Harvey delayed publishing, and instead spent his time developing his medical practice. He became personal physician to James I, who died in 1625, and later to Charles I. But he could not leave his new ideas entirely alone. He was, for example, thrilled when he was able to demonstrate to sceptical colleagues that the heart really is a pump, thanks to an unfortunate accident that had happened to the son of a royal connection, Viscount Montgomery. The Viscount's son had fallen from a horse when he was a boy, leaving a gaping hole in his chest. As Harvey wrote, "It was possible to feel and see the heart's beating through the scar tissue at the base of the hole."

He finally published in 1628, more than 12 years after his first experiments. The book was called *Exercitatio Anatomica de Motu Cordis et Sanguinis in Animalibus* (*An Anatomical Exercise on the Motion of the Heart and Blood in Animals*). It was received slightly better than he had feared and his reputation survived largely intact. Harvey had unwittingly found the first evidence for a more mechanistic view of the body, but although his breakthrough answered one question, it raised many more. What does the liver do? And what is the role of the lungs? Accepting that blood circulates meant abandoning the security and completeness of accepted wisdom.

When England erupted into Civil War, Harvey remained with the King as his personal physician, eventually retreating to the Royalist stronghold of Oxford. Here, he inspired a group of young experimentalists, including Robert Boyle, who would later become known for his observations on the pressure of trapped air (see Chapter Two), and Robert Hooke. While at Oxford, they researched blood transfusions, conception, embryology, and a wide range of other disciplines. Harvey immersed his disciples in the concept of "experimental science", and started a tradition in England that would shape scientific investigation for centuries to come. The experimental method was given a further boost when Boyle and others founded the Royal Society in 1662.

Below: An illustration of one of William Harvey's animal dissections during a lecture on the circulation of the blood.

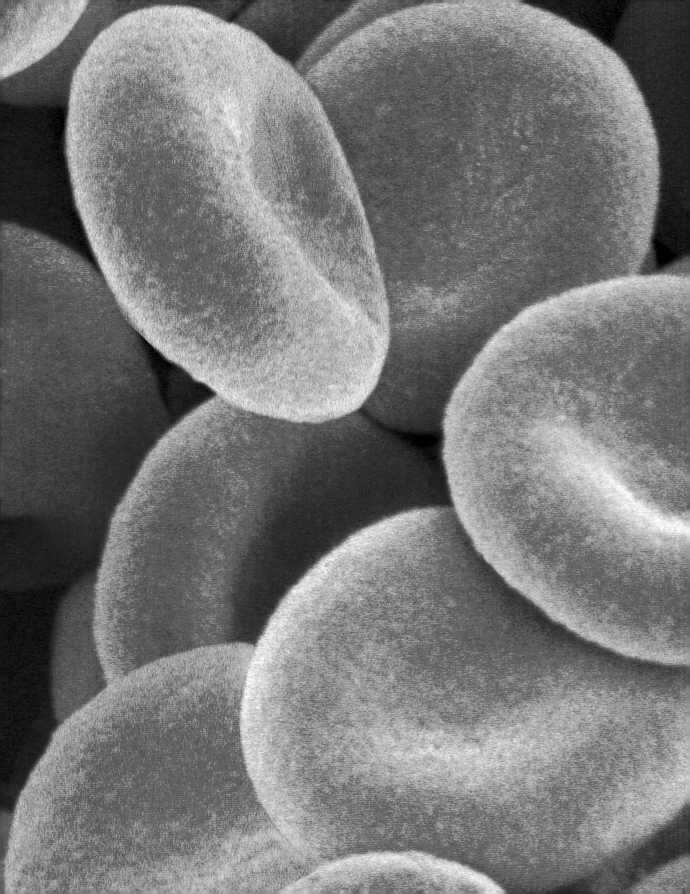

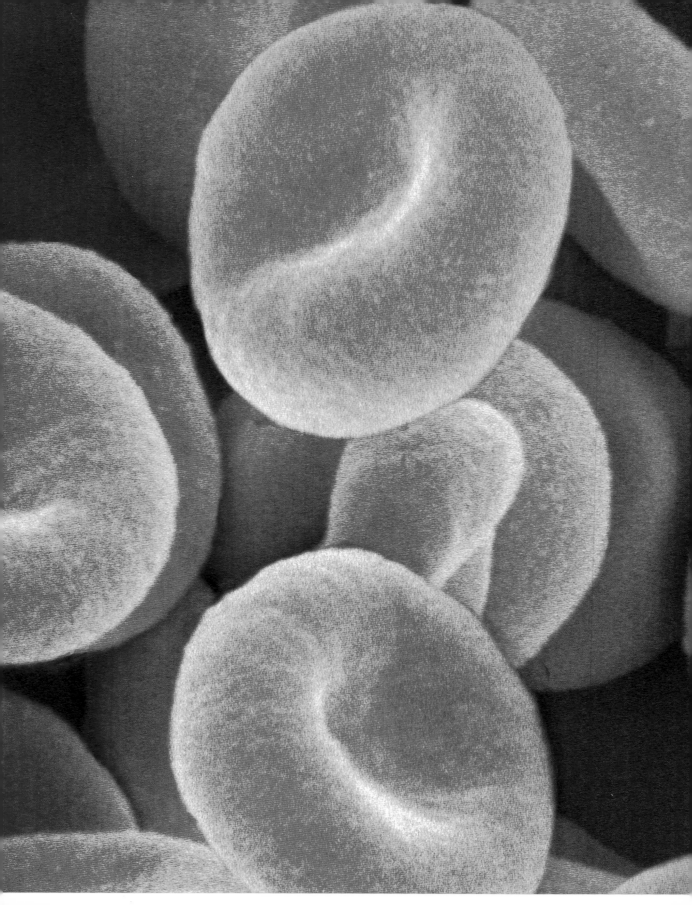

BIRTH OF CHEMISTRY

In the early 1800s, Wilhelm von Humboldt became Education Minister in Prussia and utterly transformed what was seen as the purpose of education, particularly at the university level. As a result of his sweeping reforms, universities became places of research, in which the students participated, rather than simply places where students came to learn by rote. This became the educational model on which the great American universities, such as Yale and Harvard, were founded. The emphasis was now on knowledge as a process rather than as a product. "Education" was seen as an important way of creating considerate citizens, inculcated with intellectual virtues such as self-reliance, autonomy of judgement, and critical thinking.

Humboldt's educational reforms brought with them better funding for universities and created a surge in interest in many different areas of research. Germany, for the first time, overtook Great Britain and France as the country where new technologies would be developed and discoveries made. Students in their droves began to study chemistry, travelling to other countries, learning their secrets and improving on them. This opened up another path to exploring what goes on inside a human life – a path that would be fully exploited by the Germans. Could the recently established discipline of chemistry open up the secrets of life any more effectively than electrophysics had?

An early believer that life can be explained by chemistry was Justus von Liebig. He established a research centre at Giessen in 1824 and pursued bodily functions with an alarming zeal. He made his students analyse, test, and count the constituent chemicals of everything that went into the human body – including, food, gases, and water – and, unfortunately for the students, everything that came out. He showed that the weight of the ingesta (or inputs) exactly equaled the weight of the excreta (or outputs). For him, this was evidence that life can be reduced to a series of chemical equations. Life, he decided, is not that special. Despite his meticulous analytical methods, however, Liebig's approach was criticized. The great French physiologist, Claude Bernard, said that it was like trying to work out "what occurred in a house by measuring who went in the door and what came out of the chimney". So, in the search to find the secret of life, the focus was narrowed from looking at the chemical processes of the whole body to looking for specific chemical reactions occurring inside the body.

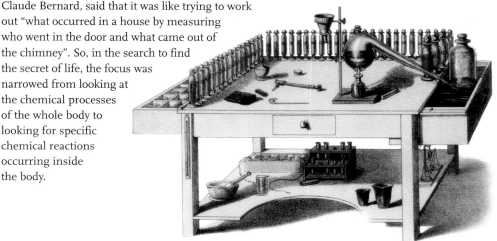

Left: Justus von Liebig 1803–1873. Liebig's investigations of the human body led him to develop important chemical processes – his inventions included artificial fertilizers and the Oxo cube.

Right: An illustration of Liebig's workbench shows the curiously shaped "Liebig condenser", still widely used in school laboratories.

HORMONE HEAVEN

In the 1880s, it was a "well-known fact" that ejaculation produced a "mental and physical debility which is in proportion to [its] frequency". For some, this was simply the first law of thermodynamics as applied to humans; energy can be neither created nor destroyed, and thus, if a body contains a certain amount of energy and semen – which can go on to produce life – then it follows that vital bodily energy must be lost when ejaculating. According to this logic, sexual abstinence maintains the energy levels of the body, which may be one reason why the Victorians were so exercised on the topic of masturbation. In the Collège de France, however, where the maxim "Not preconceived notions, but the idea of free thought" is burned in golden letters above the main hall, an elderly professor drew other conclusions. Charles-Édouard Brown-Séquard reasoned that if a person lost vitality when he lost semen, then semen must contain something vital.

**Charles-Édouard
Brown-Séquard**
1817–1894

Brown-Séquard was no medical maverick. He had been one of the first to suggest that there are chemical substances, excreted from certain organs, which travel in the blood and affect how other organs behave. In 1856, he had shown that if you remove the adrenal glands from animals they die. (It would later be discovered that the adrenal glands produce, amongst other things, adrenaline, the so-called "fight or flight" hormone, because it prepares the body to either fight or run away fast.) So, as an eminently respectable scientist, Brown-Séquard was listened to, even when describing his most controversial experiments. Aged 72, he took testicular blood from dogs and guinea pigs, added semen and juices from their crushed testicles, and injected the mixture into his own arms and legs – on ten separate occasions. He reported in the medical journal *The Lancet* that as a result of the injections his physical ageing had undergone, "a radical change". In a book he wrote about the experiment – unimaginatively titled *The Elixir of Life* – he went on to claim that following the injections he felt 30 years younger. He could now, apparently, work into the night and walk up the stairs without holding on to the banisters. And, he noted, his bowel movements had improved. Derided by some as "senile aberrations", his report nonetheless garnered interested in a new group of substances that came to be known as hormones.

It turns out that the hormone testosterone, present in dog and guinea pig testes, is critical to the development and maintenance of male muscle and mass. But Brown-Séquard's claim that injecting extract of animal testicles caused the physical changes he wrote about is ludicrous. We now know that animal tissue cannot be assimilated like that. It is more likely that what he was experiencing was the placebo effect – the triumph of hope over all too human flesh. Brown-Séquard's dog experiments were largely ridiculed by the medical establishment, but that did not stop other (male) scientists trying to find out what they could do with human testes. In 1914, George Frank Lydston gave himself a testicle transplant and in doing so, apparently, a new lease of life. "It invigorated me greatly," he proudly proclaimed. At San Quentin Prison in California, testicles were taken from executed prisoners and transplanted into others – in all, a thousand testicular transplants took place, with many reporting rejuvenating effects. Soon, celebrities were queuing up to receive bizarre treatments designed to increase the flow of testicular output, Sigmund Freud and the poet W.B. Yeats among them. Yeats spoke so often of the "second puberty" that he enjoyed, and the creative outpouring it engendered, that the Dublin press nicknamed him the "gland old man". Yeats boasted that it revived in him his "creative power", not to mention a "sexual desire" that he claimed would, in all likelihood, last him the rest of his days.

It was not until 1927 that Fred Koch of the University of Chicago finally found a way to extract testosterone from bull's testicles. He was fortunate enough to work near the enormous Chicago stockyards and able to get his hands on a regular, fresh supply. The process of extraction must have been exhausting and nauseous in the extreme; from over 18kg (40lb) of bovine testicles, he was left with just 20mg of testosterone. He then injected this testosterone into castrated pigs and rats. He discovered that when he did this they were remasculinized. By the 1930s, people had managed to make testosterone synthetically and we now know that the hormone is essential for making the male body grow. The default position of any foetus is female and unless that foetus receives a blast of testosterone between 7 and 12 weeks after conception it will become female, even if genetically male. This is why men have nipples; the nipples develop before the first testosterone blast occurs.

Right: Hormones in the human endocrine system are produced by just a few organs – the pineal and pituitary glands in the brain, the thyroid in the throat, the thymus in the chest, adrenal glands by the kidneys, the pancreas, and the ovaries or testes.

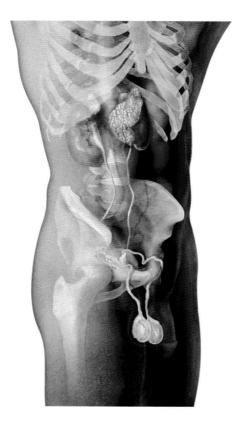
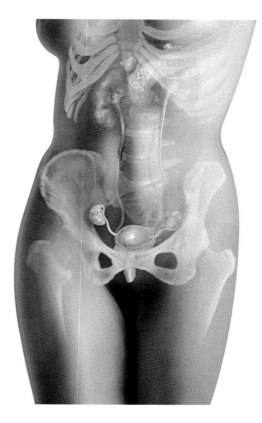

Left: Hormones are the principle signalling system responsible for triggering cell differentiation into different organs and body parts in a developing foetus.

Right: Widespread availability of the contraceptive pill from the 1960s onwards triggered a social revolution in western societies.

Soon it was discovered that hormones perform a wide range of tasks inside the human body: sex hormones determine whether a baby is male or female and help shape the brain, explaining some so-called stereotypical male/female behaviour; growth hormone controls growth; dopamine produces feelings of wellbeing and euphoria; cortisol controls the activity of the immune system; and there are many more besides. The discovery of the powerful effects exerted by hormones changed society, giving us unparalleled control over our own bodies. Steroid hormones, adrenaline, insulin, and thyroxine led to medicines that saved many lives. Even more widespread in its impact was the discovery in the 1930s that the hormone progesterone can stop a woman ovulating, leading eventually to the first oral contraceptive, approved in the US in the 1960s. Obtaining progesterone from animals was prohibitively expensive, but in 1941, an American professor from Penn State University found it could be synthesized from Mexican yams. For a surprisingly long time there was little interest in this work; there were such worries about the safety of synthetic progesterone, and whether it would lead to promiscuity, that it was not universally available to married women until 1965. Unmarried women had to wait even longer. Initially, this form of contraceptive pill had complications, so another hormone (oestrogen) was added to make the combined pill, now taken by more than 100 million women worldwide.

Nowadays, hormones are taken not only to control fertility and as replacement therapy, but also by middle-aged people in the hope of extending their youth. Hormones clearly play a hugely important role in the human body, and they do this by acting upon the many different types of cells that it is composed of.

HORMONES

The body's alternative communication system to the electrical (nervous) system is the hormone system. Among many other things, hormones determine gender, age of puberty, growth rates, and who you are attracted to. Although generally slow acting, they can also be incredibly fast – a flood of adrenaline into the body, priming it for fight or flight, will produce results within a fraction of a second of being terrified. On the other hand, it takes a lot longer for the effects of growth hormone to show. There are hundreds of different types of hormones, but broadly they all act in the same way. They are molecules, chemical messengers, that travel through the blood stream. On arrival at their destination, they dock at receptors on the surface of cells and cause those cells to respond in a huge number of different ways. Hormones are produced by tissues, such as muscle and fat, as well as in glands like the pancreas, which produces insulin (see diagram), the ovaries (oestrogen), and the testicles (testosterone). Their activities are coordinated by a part of the brain called the hypothalamus.

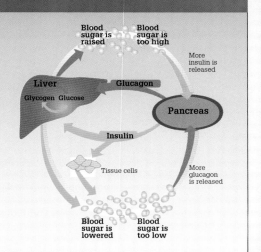

Blood sugar is raised
Blood sugar is too high
More insulin is released
Liver
Glycogen Glucose
Glucagon
Pancreas
Insulin
Tissue cells
More glucagon is released
Blood sugar is lowered
Blood sugar is too low

THE CELL

Each of us consists of over 60,000 billion cells. Each one is like a miniature chemical works, manufacturing and churning out over 200,000 different proteins at quite extraordinary speed. The cell is one of the most complicated things we know, yet each one is only a few hundredths of a millimetre across. Each time we think, move, or feel hunger or sadness, this is being mediated by cells.

Robert Hooke
1635–1703

The first person to describe and write about a cell was Robert Hooke, who we met earlier battling with Newton to demonstrate that the planets are held in their orbits according to the inverse square law (see Chapter One). He was also a noted disciple of Harvey's belief in the power of the experimental method. Perhaps Hooke's greatest work, however, was his *Micrographia*, published in 1665. It contains wonderful drawings and observations he made using microscopes – at the time, devices that were little more than sophisticated hand lenses, consisting of a piece of brass with a single hole in it containing a tiny lens. Despite this extremely limited technology, Hooke succeeded in exposing a previously unknown world. One of his many striking findings emerged when he turned his microscope on a sliced section of cork. He described the strange, regular structures he saw as being like austere monks' cells in a monastery and thus christened them "cells".

Below: This microscope, made for Hooke by instrument-maker Christopher White, allowed him to make the observations published in the *Micrographia* (left).

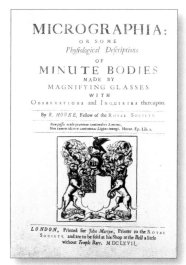

The success of *Micrographia* meant that he was widely regarded in England as the leading expert of the microscopic world, and so it was to Hooke that fellows of the Royal Society turned in 1674, when a mysterious package arrived from Holland. The package contained a letter and drawings from Antonie van Leeuwenhoek, a cloth merchant who claimed to have discovered a hidden world, a world populated by creatures too small to see with even the most powerful English microscope. This draper from a small Dutch town had created drawings of a magnification that no one had seen before – and would not be bettered for another 200 years. The fact that he was a draper was no coincidence; drapers had to inspect their cloth with magnifying lenses to assess the quality. What they were doing was counting the threads – the more threads, the higher the quality of the linen. Van Leeuwenhoek's secret was his home-built microscopes. His design was simple and not unlike Hooke's – a brass plate with a hole in it, holding a single, tiny lens – but his immense achievement was that he was able to grind the most exquisite lenses. The size of a pinhead, they were almost round; this was important for impressive magnification, because the smaller the lens and the greater the curvature, the higher the magnification.

Since he had few connections in the scientific world and spoke little English, van Leeuwenhoek had to hire a translator when he wrote to the Royal Society. In his letter he claimed to have seen tiny animals, which he called "animalcules", in pond water. He estimated that there must be around a million in each droplet of water. Fearful that he would be ridiculed for making such claims, he wrote, "I oft times hear it said that I do but tell fairytales about the little animals." The Royal Society also received detailed drawings of a bee's sting and the entire magnified body of a human louse, as well as drawings of the "animalcules" he claimed he could see swimming in his semen.

"The cell is one of the most complicated things we know, yet each one is only a few hundredths of a millimetre across. Each time we think, move, or feel hunger or sadness, this is being mediated by cells."

DNA MAKES PROTEIN

Proteins play an essential part in virtually every process within the body. They form the scaffolding that makes up the cell wall and a significant part of muscle fibre, and are an important part of the immune system, sending signals to recruit immune cells and direct their responses to disease. They are also enzymes – molecules vital to speeding up many of the chemical processes that take place within the body. Proteins are made in a complex but beautifully orchestrated sequence of events involving genes – short stretches of DNA, 30,000 of which are strung along the full length of our DNA. When the time comes for the cell to make a protein, the DNA partially unwinds. A short stretch is then copied (transcribed) into a stretch of RNA – ribonucleic acid. DNA is often described as a blueprint because it provides the template from which RNA is manufactured. Once fully formed, RNA is processed to remove the non-coding pieces and then transported out of the nucleus to the protein-making machinery (ribosomes) in the surrounding cytoplasm. Here, the code from which the RNA is made is "translated" into the appropriate protein.

ANIMALCULES

At first van Leeuwenhoek's claims were greeted with some scepticism. And when Hooke tried looking at samples of water from the River Thames, he saw nothing. "I concluded therefore that either my Microscope was not as good as the one that he made use of, or that Holland might be more proper for the production of such little creatures than England," he wrote. Hooke, however, gave the Dutchman the benefit of the doubt and persisted. Finally, after many modifications to his microscope, he managed to see what van Leeuwenhoek had seen. The creatures were blurred and in nothing like as fine detail as in the van Leeuwenhoek drawings, but they were definitely there. This draper from Holland was definitely on to something – though how he could see in such great detail remained a mystery to those at the Royal Society. The secret of how he built his microscopes was something that van Leeuwenhoek kept firmly to himself.

Antonie van Leeuwenhoek
1632–1723

Encouraged, van Leeuwenhoek sent more letters, including increasingly detailed drawings, sometimes accompanied by actual samples. He described red blood cells coursing through the veins in the tail of a tadpole – the first time these cells had been seen in a living body – and drew the capillaries linking artery to vein that only a few decades earlier Harvey had deduced must exist but could not see. He studied the texture of wood, the cells of plants, and the fine structure of animal bodies; he saw the crystals responsible for the agonies of gout; he noted the structure of nerves, muscles, bones, teeth, and hair, and examined the fine structure of 67 species of insect, 11 species of spider, and 10 of crustacean. His genius was finally recognized when in 1680 he was made a Fellow of the Society. Perhaps thinking he deserved a more prestigious name to go with his new status, it was at this point he changed his name from Leeuwenhoek to van Leeuwenhoek.

Although his observations may have astonished the Royal Society, however, their impact on the wider world was, for the time being, minimal. Medical physicians were not impressed because his findings did little to help them diagnose or treat disease any better than they had before. That microscopic life could be connected in any way to something that acted at a bodily scale seemed, to many, highly improbable. Physicians and philosophers condemned microscopy as a distraction that was irrelevant to understanding the nature of human life, health, and disease. And so the initial excitement generated by magnification evaporated. It had been hoped that the microscope would reveal the elements we are made of, but it only revealed smaller and smaller structures. Where people had thought that microscopy would provide answers, it only raised more questions. Van Leeuwenhoek's discoveries didn't spawn a tradition of microscopy because none of the great 17th-century microscopists were able to attract disciples. The secrets that the cells contained would have to remain secret a while longer.

Above left: Illustration of one of van Leeuwenhoek's microscopes – a simple design using a pin to secure the specimen and a screw to adjust the lens position, mounted on a metal frame.

By the mid 19th century there had been some progress. In the early 1800s, the botanist Robert Brown had brought back previously unknown species of orchids from an expedition to Australia. When he studied them with his microscope he noticed that the orchid cells had in their centre something that looked different – darker. He called this area, "the nucleus". He had no idea what the nucleus did but he was aware that every cell he looked at had one.

In 1839, a German scientist, Theodor Schwann, announced that animals, like plants, are made up of cells. He described animals as a "co-operative of cells", each acting independently and yet working together for the good of the whole. This was the foundation of "cell theory", one of the most important ideas in all science. But as with the discovery of "tiny animalcules", these ideas, though revolutionary in nature, had little impact on the real world. Just how little impact these insights had is well illustrated by the story of one of the great, unsung, and tragic heroes of medicine – the Hungarian Ignaz Philipp Semmelweis.

ELECTRICITY IN THE BODY

The human body has two very different communication systems – ways of sending messages from one part of the body to the other, or from one cell to another. The first is the hormone system (see page 204), which is generally rather slow, and the second is the electrical system, which is extremely fast in comparison. Electrical messages travel in our bodies along a nervous system consisting of millions of nerve cells called neurons. These neurons have fibres that can be as long as a metre and come in various guises: sensory neurons carry messages from our senses to the central nervous system and then to the brain, while motor neurons pass messages in the opposite direction. If, for example, you pick up a burning hot metal pan handle, the message will travel from the pain fibres in your finger, along the sensory neurons to the central nervous system, and before you are even aware of what has happened a message will have gone back to the fingers via the motor neurons telling you to "drop the pan". These sorts of pain messages travel extremely fast, at speeds of well over 320km per hour (200 miles per hour) – speeds comparable to a high performance racing car. Even so, this is millions of time slower than the electricity that travels along a wire.

Dendrites • Cell body • Nucleus • Axon • Stimulus • Nerve impulse • Myelin sheath cells • Axon terminal bundle • Chemical transmission

INSIDE THE CELL

By the middle of the 19th century, the spread of "germ theory" and the growing acceptance that microbes can kill us had once more focused people's attention on a world only visible through a microscope. Important things seemed to happen at the cellular level, but what was actually going on inside the cell itself?

In 1868, Dr Friedrich Miescher arrived at a castle in Tübingen in Germany to study blood. For his research he needed copious amounts of pus, a substance rich in infection-fighting white blood cells. This particular part of Germany had been at war with Prussia and there were plenty of injured soldiers lying around with weeping wounds. Miescher took pus-soaked bandages and used an enzyme called pepsin, which he got by scraping out the mucus that lines the stomachs of pigs, to break down the cells. Once the pepsin had destroyed the walls of the white cells, he was able to study the contents of the nucleus. He found, as he expected, that it contained lots of carbon, hydrogen, nitrogen, and oxygen, but he also showed that it contained phosphorus. This was surprising because he had imagined that the substance at the heart of the cell would be a protein, and proteins do not contain phosphorus. Whatever it was that he had found in the nucleus was something new. He called it nuclein, because it came from the nucleus; we know it now as deoxyribonucleic acid or DNA. Curious, Miescher began looking for nuclein in other human cells and also in the cells of a wide range of different creatures, from frogs to salmon. And wherever he looked, he found it. DNA was clearly universal and important, though Miescher himself, it seems, never realized just how important. Nor, for nearly 60 years, did anyone else.

Miescher began looking for nuclein in other human cells and also in the cells of a wide range of different creatures. And wherever he looked, he found it."

Right: An artist's impression of the structure of Miescher's "nuclein" – the famous double-helix of DNA.

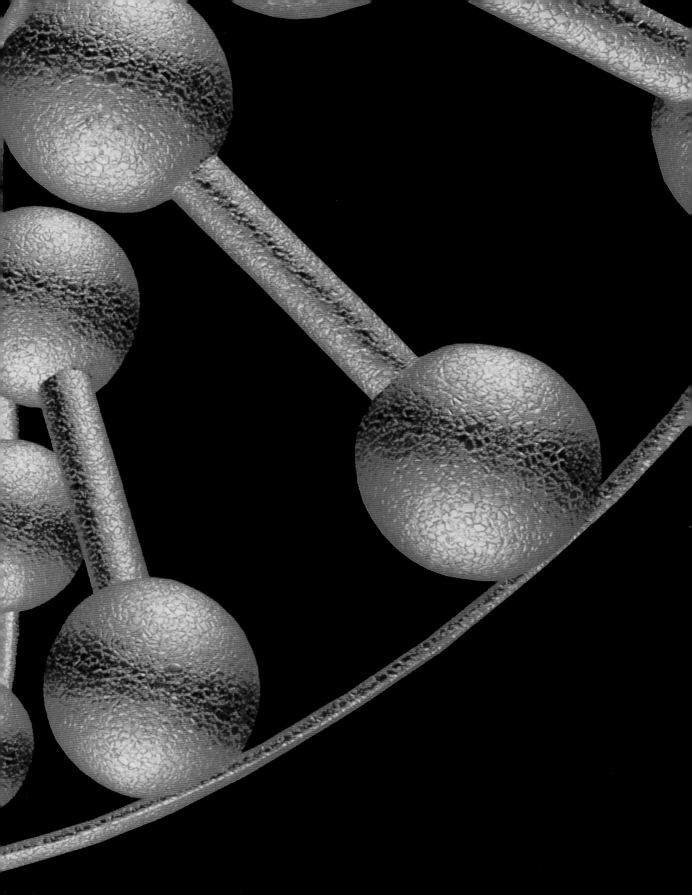

DNA COMES OF AGE

By the early 1930s, there had been massive developments in most of the sciences, particularly physics and biology, but little progress in understanding Miescher's nuclein. The nucleic acid molecule, as it was now known, had been found to contain almost equal amounts of four particular chemicals, but it seemed highly unlikely that something so simple could be significant. Certainly, it was not thought significant enough for anyone to devote a great deal of their time to studying it.

It was Fred Griffith, one of many scientists of the time who were not particularly interested in DNA, who made the next major discovery. Griffith, a microbiologist working in London, was interested in pneumonia, and in particular the bacteria that cause it. He had discovered that there are two different strains; one that is rapidly lethal and one that is harmless. When he killed the lethal strain by heating the bacteria and injected them into mice, the mice, unsurprisingly, survived unharmed. In 1928, however, he made a discovery that was utterly unexpected: if he killed some of the lethal strain, as had before, and then mixed the bacteria with the harmless strain, something rather disturbing happened – the live, "harmless" bacteria now sometimes became lethal, capable of killing mice it was injected into. It seemed that some unknown substance, which could not be killed by heat, was being taken up by the previously harmless bacteria and transforming them into killers.

Griffith did not pursue this line of research and was himself killed during an air raid in World War II, well before the implications of what he had achieved became clear. Instead, his findings were taken up by a Canadian called Oswald Avery working in New York. Avery, like Griffith, had spent many years studying the bacteria that cause pneumonia. When he heard what Griffith had done, he switched his area of research and spent many years obsessively testing every possible component of the bacteria to discover what it was that was causing the change. He removed the lipids, then the carbohydrates, then the proteins. It was none of these. Finally, almost reluctantly, he turned his attention to DNA, the least likely of the potential candidates. He discovered that if he removed their DNA the bacteria lost their ability to kill mice; they were no longer lethal. In 1944, 16 years after Griffith's original research, Avery finally felt confident enough to publish his findings. His progress was slow, partly because he was thorough and partly because he was opposed every step of the way by his boss, a man who thought Avery was wasting time on something that would turn out to be unimportant. Despite the fact that it soon became clear Avery had made a hugely important discovery, his boss lobbied the Nobel Prize committee to make sure that Avery never received his reward.

"He switched his area of research and spent many years obsessively testing every possible component of the bacteria."

Left: Oswald Avery 1877–1955. Canadian-born Avery spent most of his life in New York. He has been described as the most deserving scientist never to be awarded a Nobel Prize.

WHAT IS DNA?

By the end of World War II things were finally becoming clearer. Every living thing is made up of cells. These cells, it had emerged, contain genes, which determine how the organism grows and behaves. The genes are mainly located inside the nucleus and are made of DNA. Yet still, what DNA is and how it manages to do what it does remained largely a mystery. Part of the mystery, however, was soon to be solved.

World War II had been not only a war between armies but also one between scientists. During the war, many of the best minds had been drafted into working on ingenious ways to kill the enemy, the most complex and deadly being, of course, the atomic bomb. The bomb was the result of developments in atomic theory that had begun at the start of the 19th century (see Chapter Two). One of the other products of that research was quantum theory, an apparently esoteric branch of physics that would prove to be central to the development of the electronic age, but also to explaining the chemical bonds that define the structure of DNA.

After the war there were many disillusioned scientists who wanted to work on something more edifying, more life affirming, than bombs and other ways of killing their fellow human beings. So physicists like Maurice Wilkins and Francis Crick turned to biology. Wilkins, a shy New Zealander, went to work at Kings College in London. Here he had access to the latest X-ray imaging techniques, techniques which were allowing scientists to look in much finer detail at the structure of matter. He was joined in the early 1950s by a brilliant young researcher called Rosalind Franklin. She was put to work deep in the basement, studying DNA, and in the course of her work was exposed to huge numbers of X-rays, which may have contributed to her tragically early death. Franklin got images of the structure of DNA by teasing out single strands and then exposing them to X-rays. The X-rays scattered as they passed through the DNA and the results were captured on a photographic plate. In all, Franklin took over 100 pictures, each of which could take up to 90 hours to produce. She was painstaking and meticulous in her research, but did not get the credit she deserved. Instead the glory went to James Watson and Francis Crick.

> *"Franklin got images of the structure of DNA by teasing out single strands and then exposing them to X-rays."*

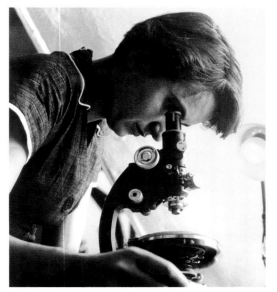

Left: Rosalind Franklin 1920–1958. As well as her important work on the structure of DNA, Franklin investigated the tobacco mosaic virus and polio viruses.

Right: The distinctive pattern on this X-ray crystallography image revealed that DNA had a helical shape similar to a spiral staircase.

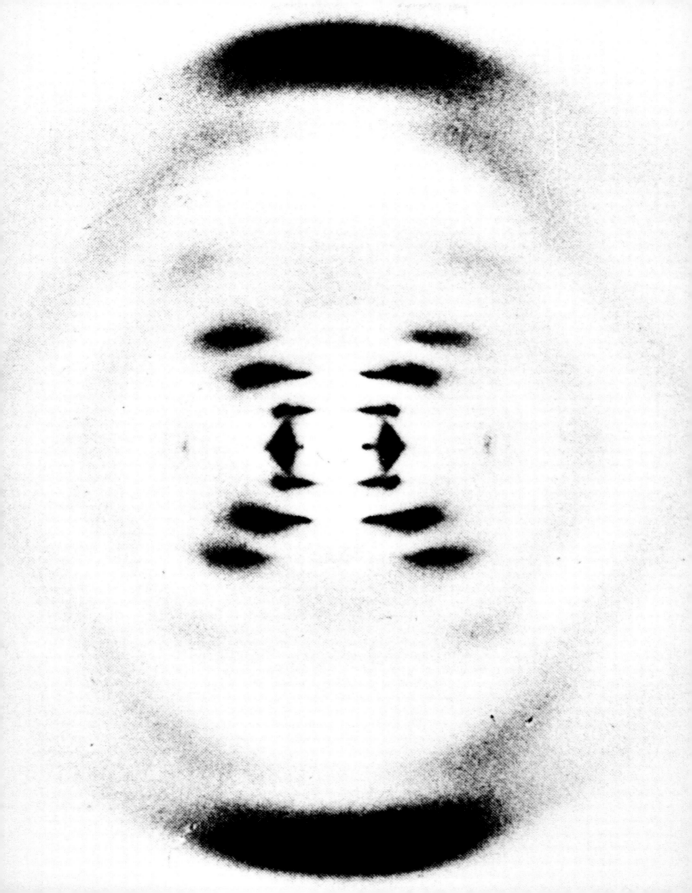

THE SECRET OF LIFE

In the 1950s, Cambridge University was a hub of intellectual research – incredibly bright people who were happy to share their ideas were always coming and going. This was one reason why both Crick and Watson chose to work there, and it was a smart decision, because it was thanks to certain chance encounters that they were able to deduce their famous structure. Although they had realized DNA must have a helical structure, they did not understand how all the parts fitted together; it was only when Watson met with Maurice Wilkins and was shown some of Franklin's images (without her knowledge or consent) that they were able to create a realistic model. Legend has it that on 28 February 1953 Crick strolled into the Eagle pub in Cambridge and announced, "We have found the secret of life." News of their discovery made its way back to London where Rosalind Franklin was preparing to send her own manuscript to the science journal, *Nature*. In the end she published, alongside Watson and Crick, in the 25 April issue. What she did not know was that Watson and Crick's model was actually based on her data. She died five years later from cancer, four years before the Nobel Prize for this particular work was awarded to Watson, Crick, and Wilkins.

Watson and Crick realized that DNA consists of two long strands, which are intertwined in the shape of counter-rotating spirals. On the sides that face each other – the insides of the spirals – are just four molecules: adenine (A), guanine (G), cytosine (C), and thymine (T). Each human cell contains about 3.4 billion of these "letters". If DNA is thought of as "The Book of Life", as it is often referred to, then it comes packaged up in chromosomes ("chapters") and broken down into genes ("paragraphs"). The wonderful simplicity of DNA is that every letter on one strand is always matched with a corresponding letter on the other strand. C always pairs with G and T always pairs with A. So a small section of a strand of DNA which reads AACGGTCA will be matched with a strand that reads TTGCCAGT. The beauty of this is that it explains exactly what happens when a cell divides: the DNA separates into two different strands and, for each, a new partner strand is built, which entwines with the original strand. Another way to understand the process is to imagine two dancers – one male and one female – in close embrace. They briefly separate and in that instant an identical copy of their partner is created, with whom they each now happily entwine.

> *"This is how living things grow and repair themselves, simply by producing new cells containing identical strands of DNA."*

Left: James Watson (left) and Francis Crick with their groundbreaking 1953 model of the DNA double-helix structure. Phosphate "backbones" running along each side are linked by "base pairs" forming the "rungs" of a spiral ladder.

This is how living things grow and repair themselves, simply by producing new cells containing identical strands of DNA. Hair, for example, grows because the roots are hard at work churning out billions of identical new hair cells. If, however, I cut myself and need a quick repair job it is rather more complicated. My body responds to the damage by producing copies of lots of very many different types of cells (skin cells, white blood cells, platelets, and so on) to cope with the damage. Since every cell contains identical DNA, this does beg the question of how exactly the cells "know" what to do and what their function is. It is a question that science is still wrestling with.

Although all DNA conforms to the basic "double helix" design, this complex molecule can take on many different detailed structures of "conformations". The three conformations shown here are know as A-, B-, and Z-DNA. The B-DNA form is by far the most common in living cells.

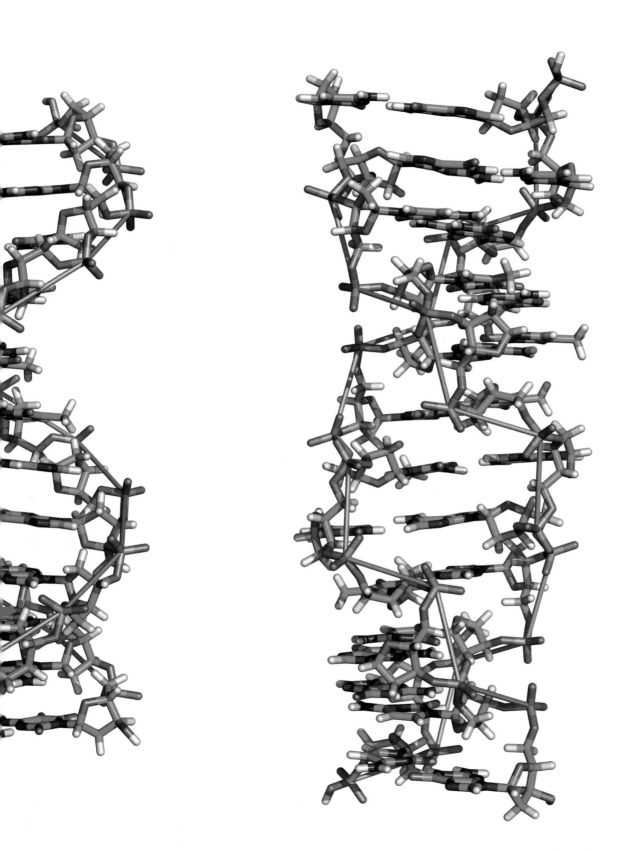

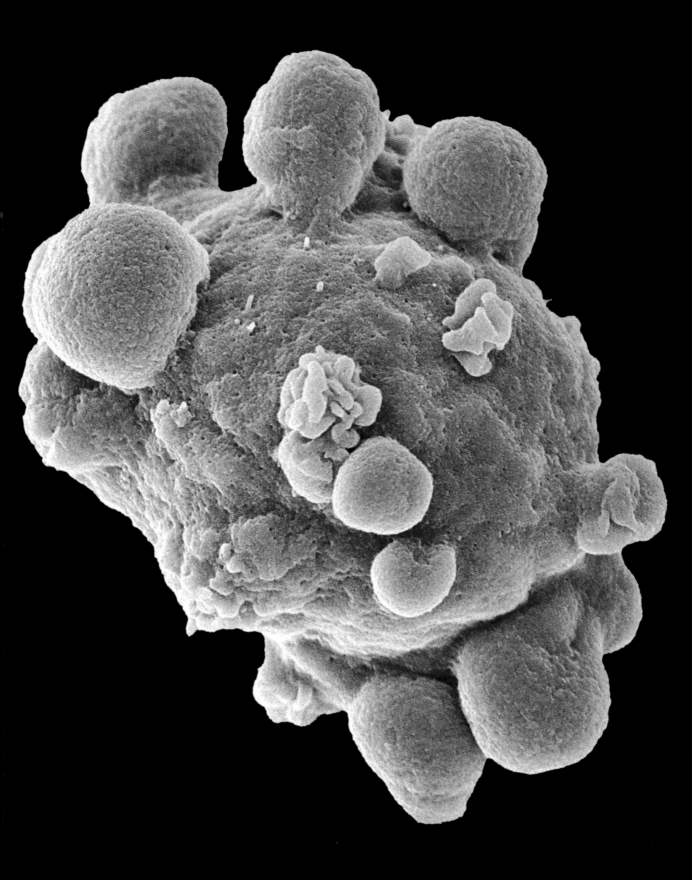

The discovery of the structure of DNA was a great moment in the history of science and in the search to discover the secret of life; understanding the structure answered some hugely important questions about replication, mutation, and human evolution. Since 1953 there have been further tremendous advances in understanding exactly how DNA replicates, how it makes proteins, how it can go wrong, and how it is that such a small number of genes (only about 30,000) can produce something as complicated as a human being. In the light of what we have learnt in just half a century it is tempting to claim, as some enthusiasts do, that we are on the brink of being able to truly control and manipulate life. The problem is that, as scientists discovered in the past, the closer we look the more complicated it gets. But although we may not yet know exactly what it is that makes us tick, we do know that one extraordinary molecule is at the heart of us and every other life form on this planet.

MUTATION

Mutation sounds terrible, like the unfortunate product of a dangerous experiment leading to the creation of some freakish or unfortunate monster. Yet if it was not for mutation we would not exist because our remote ancestors would never have emerged from the primeval ooze; favourable mutations play a crucial role in evolution, providing a source of variation on which natural selection acts. A mutation can occur when a cell divides and the letters in a stretch of DNA that forms a particular gene are not faithfully copied. A stretch of DNA reading AACCCG, for instance, could be copied as AGCCCG. When this stretch of DNA is later used to make a protein, that protein will be different to proteins produced from the original sequence. This may or may not be important. It may lead to the death of the animal, or just a subtle change, or no change at all. As a simple example, imagine a mutation changes the colour of a moth's wings from light to dark. If the moth lives on light-coloured trees then this mutation will make the moth stand out. Moth and mutation will rapidly become extinct. But if the trees become blackened – perhaps because of soot from a nearby factory – then the dark moths will blend better into this new environment. The mutation is favoured. We will see far more dark-coloured moths. Mutation has led to evolution.

Connections – Body

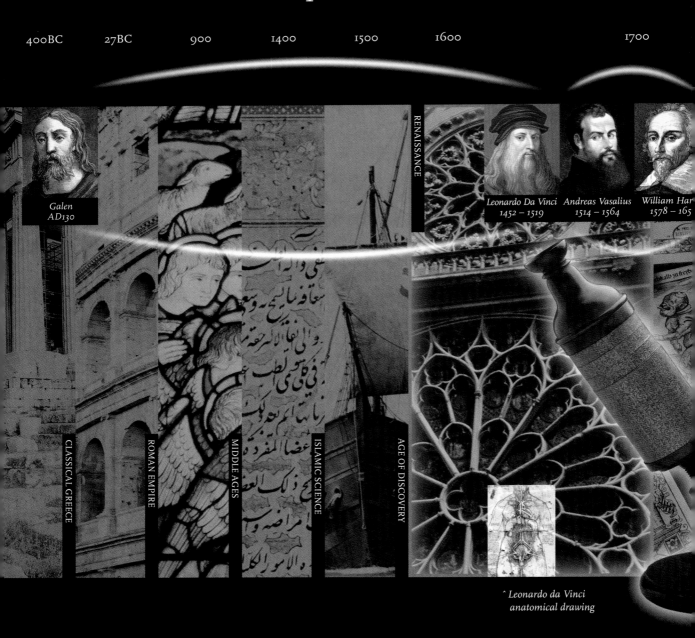

Galen
AD130

RENAISSANCE

Leonardo Da Vinci   *Andreas Vasalius*   *William Har*
1452 – 1519   *1514 – 1564*   *1578 – 165*

CLASSICAL GREECE

ROMAN EMPIRE

MIDDLE AGES

ISLAMIC SCIENCE

AGE OF DISCOVERY

^ *Leonardo da Vinci*
anatomical drawing

Our attempts to understand and prolong life have drawn us into ever more detailed inspections of the workings of the body. Unfortunately, the view of our anatomy that held sway until the Renaissance was based on dissections of animals by a Greek called Galen. The Church discouraged dissecting bodies, stifling further research. Leonardo da Vinci's incredibly accurate drawings were never seen widely; it was only when Andreas Vesalius, a professor of anatomy at Padua, bravely started an unprecedented programme of dissections that a reasonably accurate anatomical guide was created. Fifty-four years later a young British doctor, William Harvey, discovered the circulation of the blood, further discrediting Galen's work.

Antonie van Leeuwenhoek 1632 – 1723

Robert Hooke 1635 – 1703

Rosalind Franklin 1920 – 1958

James Watson 1928 –

Francis Crick 1916 – 2004

AGE OF ENLIGHTENMENT

EARLY 20TH CENTURY

MID 20TH CENTURY

21ST CENTURY

< *Robert Hookes' microscope*

^ *DNA double helix structure*
^ *DNA x-ray*

Our understanding of the body – and how to protect it – took on a new dimension in the 17th century with the use of microscopes. Robert Hooke was able to observe the cell, the building block of life itself; Antonie van Leeuwenhoek's far more powerful microscope was able to see microbes, "animalcules", paving the way for germ theory and the work of scientists like Louis Pasteur in preventing infectious disease.

From the late 19th century our focus zoomed in further still, to what made cells work. By the mid 20th century, it was clear that genes were crucial and that they were made of mysterious substance, DNA. Now, using x-rays rather than telescopes, Rosalind Franklin was able to photograph individual strands of DNA, from which Crick and Watson defined its structure, answering many questions about the body, but posing still more

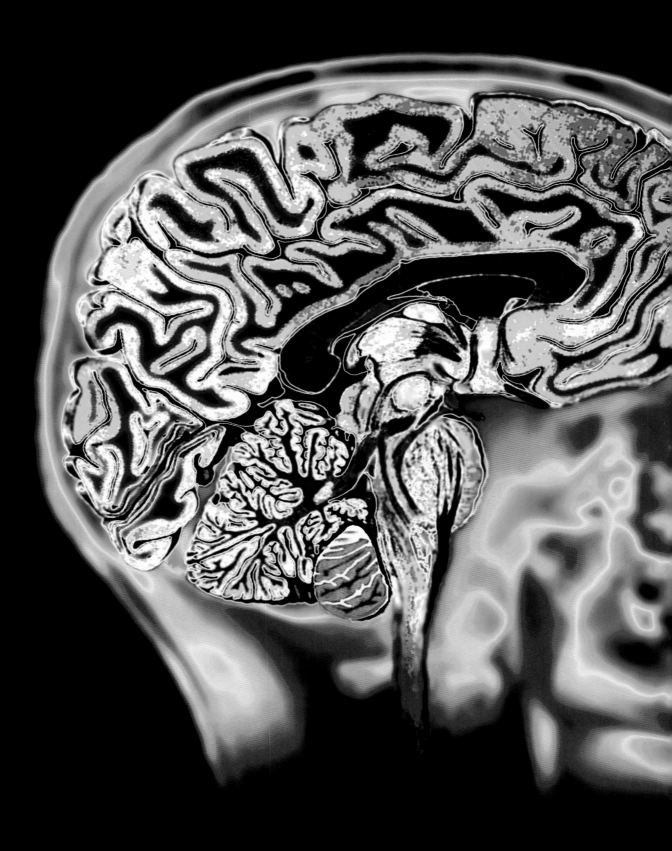

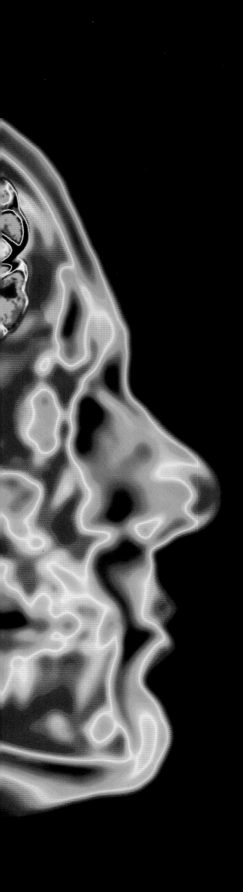

Mind

WHO ARE WE?

If you were asked to describe yourself, you would probably start with your physical appearance. But when it comes to describing precisely what it is that drives and motivates us, most of us would struggle. The truth is that few of us really understand the workings of our own mind; we often behave irrationally, procrastinate, and do unpredictable things for reasons that are not obvious even to ourselves. The search to understand who we are and what really motivates us has been a long one. Early human civilizations had little idea that the brain was responsible for cognition at all, and it has only been in the last century that we have begun to understand how the 1.5kg (3lb) of grey and white matter that sits on top of our spinal cord allows us to think. But neuroscientists are now unearthing increasing evidence that much of what the brain does lies beneath our conscious awareness and that many of the decisions we think we make are actually rationalizations – the conscious part of us justifying decisions that have, in fact, already been taken by the subconscious parts.

Our brains, like our bodies, are the product of an extremely long process of evolution. On top of a 500 million-year-old reptile brain, containing the brain stem and cerebellum, is grafted a much younger addition: the area of the brain called the neocortex, which is responsible for language and abstract thought. The neocortex, Latin for "new bark", is a thin layer of grey and white matter that sits on top of the cerebral hemispheres – the left and right sides of the brain. It is the seat of rational thought, and is seen by some as a sort of "command centre". But the neocortex is actually more like the mahout who perches on top of the elephant; sometimes he can direct the elephant along the path that he wants the elephant to take, but often the elephant goes where it wants to go and the mahout is left pretending that, yes, that is also where he wanted to go.

Left: A magnetic resonance imaging (MRI) scan of a human skull reveals the complex folds and details of the brain – the body's most complex and intriguing organ.

THINK LIKE AN EGYPTIAN

On an interminably hot day in Luxor, central Egypt, a local consular agent named Mustapha Aga made his way to a meeting with one Edwin Smith. Having arrived there from Connecticut in 1858, some four years before, Smith was an adventurer of dubious reputation. He had studied Egyptology in Paris and London and was now earning a healthy living lending money, dealing in antiquities, and, rumour had it, making forgeries. Aga was confident that the ancient scroll in his possession would intrigue the avaricious American. Sure enough, Smith was immediately interested in what Aga carefully unrolled in front of him. Examining the 469 lines of hieratic script – a less pictorial adaptation of hieroglyphics – he recognized it as some sort of medical document and decided to buy it for his own collection.

Although Smith was a scholar of sorts he never got around to translating the papyrus and for decades it remained little more than a curio. After his death, the scroll was given to the New York Historical Society and was finally translated between the world wars. It became evident that the document, thought to date from 1500 BC, was a copy of a much more ancient work, probably conceived at least a millennium earlier. It contained detailed medical case studies of 48 wounded individuals, the majority of whom had suffered head injuries of the sort that could have been picked up in battle, or, today, might have been a result of accidents on building sites. Some, it is thought, were sustained during the construction of the pyramids. In one case the author instructed: "If you examine a man with a gaping wound in his head, which reaches the bone, smashing through the skull and breaking open the brain, you should feel his wound. Like the corrugations which appear on molten copper in the crucible, something therein throbs and flutters under your fingers like the weak place in the crown of the head of a child when it has not become whole."

What had been translated was not only a textbook of ancient medical knowledge but also the earliest known descriptions of the brain. The human brain, now considered the most complex organism in the known Universe, was, over four thousand years ago, only just beginning to be understood. The Egyptians knew that the brain was covered by membranes (now known as the meninges); they knew of various ways – some sensible, some less so – to treat head injuries; and they knew of the curious paralysing effects that brain trauma could have on apparently unconnected limbs. However, despite their remarkable first inroads into the world of neurology, the Egyptians had a very different view of the importance of the brain. They thought that the soul – a person's cognitive and immortal self – had nothing whatsoever to do with the brain. They believed that the heart was the organ from which consciousness stemmed. Accordingly, while Egyptian corpses were carefully preserved or mummified for the afterlife, and the heart was pickled in a special canopic jar, the brain was scooped out with a metal hook and unceremoniously thrown away.

Right: A page from the Edwin Smith papyrus, the world's oldest surviving surgical document. The script, called hieratic, is a hand-written alternative to the hieroglyphs used on monuments.

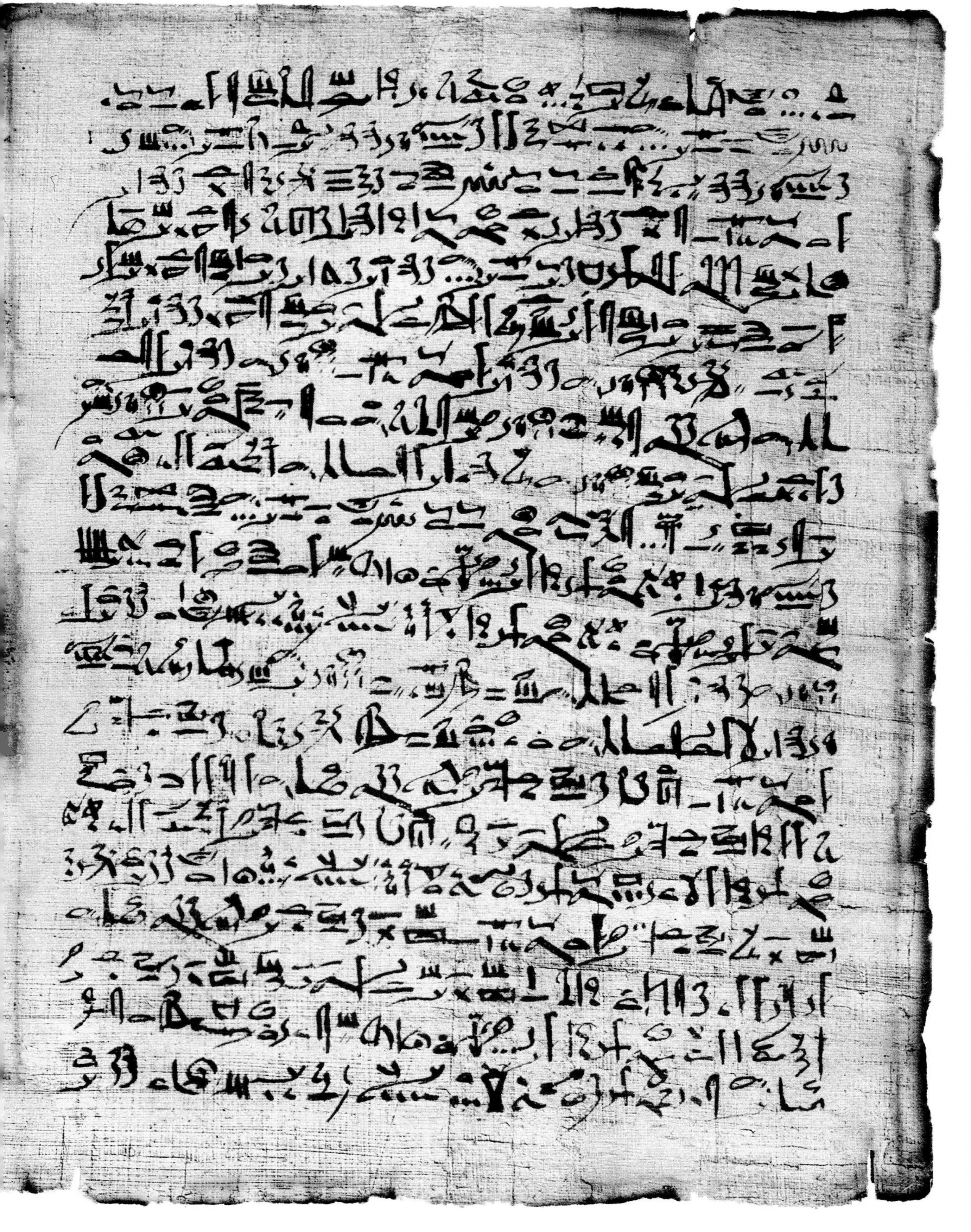

THE BRAIN COMES TO LIFE

When asked who they really are, most people will try to describe their behavioural characteristics or way of thinking – today, we are beginning to see our true selves as the workings of our mind. But the first people to seriously consider who they were in secular as well as religious terms were the classical Greeks. Hippocrates, a famous physician, not only worked out that injuries to the brain affected opposite sides of the body but also understood that epilepsy was a brain disease, not possession by an evil spirit. And while the Greeks believed that their destinies were determined by mythical beings called "Fates", they thought that day to day behaviour was controlled by an individual's psyche.

Hippocrates
c.460–c.377 BC

Plato, who lived in the 4th and 5th centuries BC, was one of the first to develop the idea that the self, responsible for our rational decisions, our reason, and our thoughts, was associated with the brain. Plato's theory, based more on philosophy than clinical examination, was that our cognitive soul, which would pass to another body after death, was made up of extremely fast-moving spherical particles, which concentrated in the head and nervous system. He suggested that other organs had psyches too, although these were not immortal. One psyche, associated with desire or animalistic appetites, inhabited the diaphragm from where it could easily energize the nearby liver, believed to be the organ of lust. Another – the psyche that controlled emotions – was seated in the heart. (The use of hearts on Valentine's cards is a throwback to these ideas.)

Right: Plato was famous for his mathematical and philosophical theories. His ideas about the mind were derived from idealized precepts rather than careful observation.

Below: A medieval Italian relief showing Plato and Aristotle – two great philosophers whose ideas on the mind proved sadly mistaken.

Aristotle, who had first learnt about the human body from his father, a personal doctor to the King of Macedonia, jumped on the bandwagon of physiological speculation. Taking a great leap backwards, he taught the civilized world that the heart was the seat of the rational soul, while preaching that the brain was little more than a bloodless radiator that allowed the body to cool off. Strokes, in his opinion, were blockages in the head caused by a build-up of black bile. However, his idea of lasting value was that there was a mysterious, almost heavenly, substance that the centralized psyche could send to the muscles to make them function. Of course, he had no knowledge of electrical impulses being conveyed along neurons or of chemical transmitters crossing synapses, but his concept pointed future generations in the right direction. Aristotle thought that all empty spaces contained a new and powerful – but invisible – fifth element called aether. This was taken into the body by the lungs and was then transformed into "vital pneuma" or "vital heat" in the heart, the organ of perception from which he supposed all the nerves emanated. The vital pneuma, a life- giving force, was then carried by the blood into muscles where it would activate their psyches, spurring them into action. Thus, what Aristotle understood was that there was one organ controlling all the others. He simply had not worked out that it was the brain.

PROBING THE DEAD...
AND THE LIVING

One reason that the classical Greeks failed to advance neurology beyond their ingenious but speculative philosophies was that they had beliefs about treatment of the dead that severely restricted what they might have learned by carrying out dissections. They thought that if a corpse was not properly attended to – in a quick and respectful burial – a person could spend eternity wandering the bleak banks of the River Styx. For this reason, human autopsies were strictly illegal. Even animal dissections were not encouraged as many classical Greeks believed in reincarnation. But with the arrival of the Hellenistic Age, heralded by Alexander the Great and the founding of his capital in Egypt rather than on the Greek mainland, religious ideas began to change. The soul and the body were no longer seen as being so intimately connected and pathologists were able to sharpen their scalpels.

Herophilus of Chalcedon, a Turkish Greek living in Alexandria, performed hundreds of human dissections and in doing so made the first detailed examinations of our brains. Slicing them into sections, he noted the four hollows (ventricles) within them, separated the two membranes that cover them, described their two major parts (cerebellum and cerebrum), and traced what he recognized as motor or sensory nerves extending from the brain stem to other regions of the body. While he made some interesting discoveries, however, his methods were exceptionally brutal. Ptolemy, the King of Egypt at the time, was so keen to explore the human brain that he presented Herophilus with condemned criminals on whom he could perform his horrific vivisections.

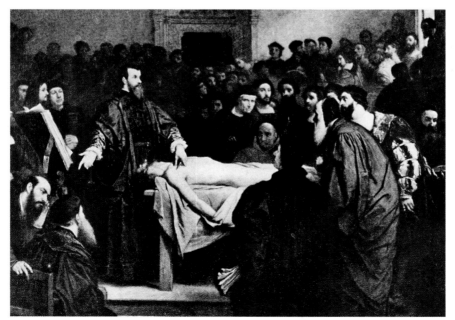

Left: Herophilus conducts a dissection before watching students in his medical school at Alexandria.

Above right: A medieval illustration of Herophilus and Erasistratus, his apprentice at Alexandria who continued the early tradition of studying anatomy through dissection.

IOCLESSHEROPHIL9ERASISTRAT9

TREPANNING AND EARLIEST BRAIN SURGERY

Although the study of the brain is still in relative infancy surgeons have been successfully opening the cranium for over 7,000 years. Trepanning, trephining, or craniotomy is a practice that involves cutting out a section of skull and letting it heal over. It was first used in Neolothic Europe well before the creation of Stonehenge. Skulls discovered at Ensisheim in France display neat apertures, often several inches across, opened up with flint blades long before their owners' deaths. Other early cultures, stretching from China to Central America, were also avid trepanners. Although not strictly brain surgery – piercing the organ's outermost layer, the dura mater, would almost certainly have resulted in death from infection – trepanning could save victims of a head wound suffering from build-up of fluid. By the 15th century, itinerant barber-doctors roamed Europe, opening the skulls of fee-paying patients after claiming they could cure their depression or epilepsy. Some even spread the myth of an extractable "stone of madness", presumably palming a pebble after the trepanation. Even today, a more sophisticated form of the practice is not only used by surgeons to relieve swellings beneath the skull but also attracts a cult following among a self-practicing New Age community who believe it can enhance well-being.

INTO THE DARK

In the previous chapter we encountered Galen, the Roman whose intellectual claims dominated Western medicine for over a thousand years. And his investigations and overbearing authority were no less influential on the question of who we are; as the physician to gladiators in the Ancient Greek city of Pergamon, he studied numerous head wounds and was able to observe their effects on the more unfortunate, or perhaps less talented, fighters. He noticed that certain bodily functions could be hindered by blows to the skull while others would remain utterly unaffected. Although he did not go so far as to try to map the areas of the brain and relate them to our different abilities, he did sow the seeds for the theory that was to be known as localization, to which we will return later.

Unlike the earlier Greek philosophers, Galen worked in a fashion that was basically empirical. Not only did he come up with theories, he also set about testing them – he experimented. Having theorized, for example, about the reasons why a sword blow to a gladiator's neck had paralyzed him but allowed him to continue breathing he would cut, in a comparable place, the necks of whichever unfortunate animals he could lay his hands on. The results of his slicing were then carefully noted down and compared. While these experiments seemed to finally prove that something in our heads controlled the rest of the body, Galen somehow failed to work out that this "something" was the brain itself. He thought that it was the spaces within it. Having lopped the tops off skulls of living animals he noted how the cerebrum beneath would rise and fall, like a lung. He concluded that air containing Aristotle's life-giving pneuma must be sucked into the front ventricle through a plate above the nose. There it would be transformed into vital or "animal" spirits, which would then be processed through the posterior ventricle before passing into the nerves and beyond, to every region of the body that required their life-giving powers. Fanciful as the idea may have been, this was to become the theory that medical students were taught for the next thousand years.

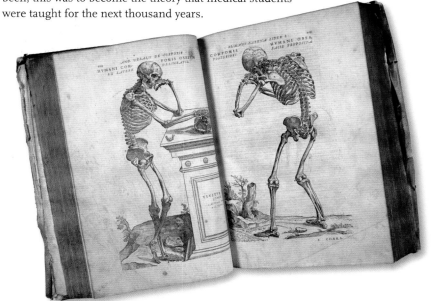

Left: Two illustrations from the work of Andreas Vesalius, whose anatomical studies in the 16th century finally overturned many of Galen's mistaken ideas.

After Galen, Christianity stalled philosophical and neurological progress across Europe. Throughout the Dark Ages, the question of our own existence was not one that was widely grappled with. It was the responsibility of the Church to make sure that everyone knew their importance – or lack of it – and accepted their place in the world as determined by an invisible God whom they had no right to question. The spirit of rational enquiry was not encouraged in this superstitious feudal society. By comparison, Islamic medicine at this time was more advanced, but the Muslim world's understanding of the workings of the brain, similarly based on the ideas of the 4th-century philosophers of Alexandria, was little different. Ali al-Husain ibn Abdullah ibn Sina, better known as Avicenna, like his European counterparts believed that the ventricles of the brain were home to our cognition. According to his theory, the front ventricle received incoming spirits and the posterior, through which animal spirits flowed to the nerves, housed memory. Reason, or the rational soul, inhabited the middle ventricles.

> *"Galen noticed that certain bodily functions could be hindered by blows to the skull while others would remain utterly unaffected."*

WITCHES, ERGOT AND LSD

In the town of Salem, Massachusetts, in the winter of 1692, eight young girls started suffering from extraordinary symptoms. They fell into delirious trance-like states and felt insects crawling across their limbs; their speech became garbled, their bodies convulsed, and they became convinced they had been cursed by a malevolent spirit. Six local women were eventually hung for witchcraft. Just over 250 years later, in 1943, the Swiss industrial chemist Albert Hoffman became the first man to ingest lysergic acid diethylamide, better known today as LSD. Within a few minutes of having accidently done so he became terrified of the furniture, incapable of riding his bicycle, and convinced that he had been invaded by a demon, and that his concerned neighbour was an evil witch. The similarities between Hoffman's experiences and those of the Salem girls are now believed to be anything but a coincidence. Hoffman synthesized LSD from ergot, a poisonous mould found on wheat. Re-examination of the climatic record of the Middle Ages points to the peaks of supernatural possession – manifested by burnings of unfortunate women and the occurrence of St Anthony's Fire, a phenomenon in which whole villages simultaneously become mentally unstable – coinciding with the likelihood of ergot infestations. Although the manner in which serotonin-blocking drugs such as LSD affect the brain is not entirely known, it is clear that such chemicals can completely change who we are.

Above: Ears of wheat infected with the black, protruding ergot fungus.

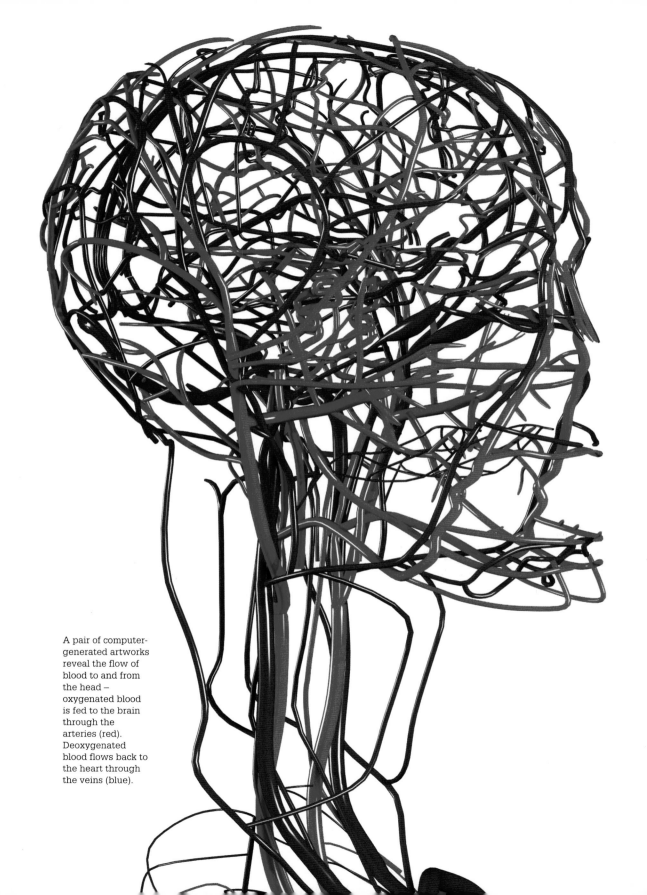

A pair of computer-generated artworks reveal the flow of blood to and from the head — oxygenated blood is fed to the brain through the arteries (red). Deoxygenated blood flows back to the heart through the veins (blue).

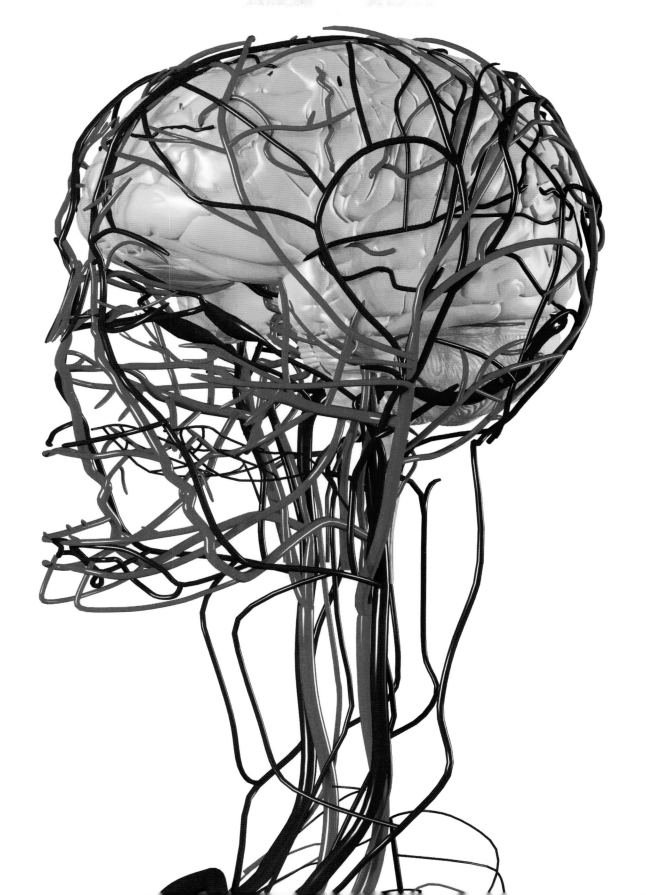

A NEW HOPE

Unsurprisingly, the Renaissance saw a transformation in the way we viewed ourselves. As Vesalius got to grips with the corpses of Padua and tore up Galen's rule book (see Chapter Five), he reopened the door of enquiry into the brain. His suspicions were aroused when he noted that the ventricles in animal brains are nearly identical to those in humans. Since the ventricles were thought to be the houses of reason and consciousness, and the idea of animals having souls was frankly preposterous, he knew something was wrong. During his search for answers, Vesalius studied the rete mirabile, a network of blood vessels at the base of the brain that had been decreed the source of the all important psychic pneuma. After probing enough skulls he realized that while the rete mirabile was a major feature of cows' brains it simply did not exist in humans. It was becoming very clear that a better model of the mind was badly needed.

René Descartes, born in a small French town in 1596, was the unlikely saviour of neurology. He was not a physician but a philosopher with a passion for mathematics and a degree in law. By his early twenties he had fought for both the Protestant Dutch army against the Spanish and for the Catholic Maximilian the Great in Bavaria. He had wandered Europe, discovered analytic geometry, and now he was bored. As he strolled and contemplated his future in a park near his home in Paris, he was enthralled by a set of extraordinary mechanical sculptures that had recently been erected. Arranged in six grottos were tableaus from classical mythology: life-sized gods and goddesses who were actually able to move. A series of hydraulic tubes laced their hinged bodies allowing them to seemingly come alive. Even their faces were able to

Rene Descartes
1596–1650

Above: An illustration from Descartes' *De Homine* demonstrates his theory that the pineal gland was key to interpreting signals from the outside world and triggering actions in response.

react to passers-by, triggered by levers buried under the surrounding paving stones. As Descartes approached one of the eerily lit caverns, the naked goddess Diana hurriedly hid herself in the reeds, Neptune surfed angrily towards him brandishing his trident from the top of a giant clam shell, and a scaly sea monster appeared from below the waves and squirted water in his face. After his initial astonishment Descartes started thinking: perhaps the levers that set their limbs in motion were in some way similar to the mechanism by which our mind controls our body? He was soon spending his time in slaughterhouses and mortuaries attempting to work out what it was that made us different to machines. As a staunch Catholic in a new world in which the Earth was no longer the centre of the Universe and the heart was no more than a mechanical pump, he needed to reconcile his spiritual and scientific beliefs. He had to work out what made us consequential. If the brain is no more complex than a complex pneumatic statue, he thought, then what makes humans special?

"Descartes had a wonderful new way of looking at the world that separated the soul and the intelligent mind from the rest of the body – dualism."

Descartes had a trick up his sleeve: a wonderful new way of looking at the world that separated the soul and the intelligent mind from the rest of the body – dualism. His idea was that the soul inhabited a tiny part of the brain called the pineal gland, a small structure shaped and sized like the pine nut from which it got its name. He decided that it was from this command centre, hanging like a punch bag near the ventricles, that our bodies are controlled. The concept can be most easily imagined by thinking in terms of a modern joystick on a computer game or on the controls of a fighter jet.

Right: The Chinese Taoist "Tajitu" symbol is a classic representation of dualism – how the yin and yang – body and mind, dark and light, male and female – come together to form a whole.

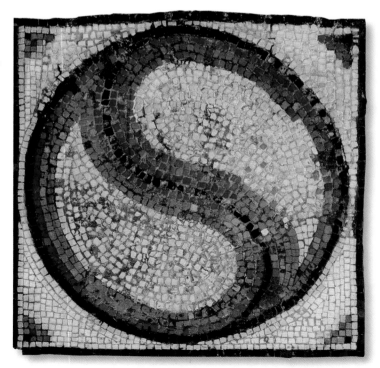

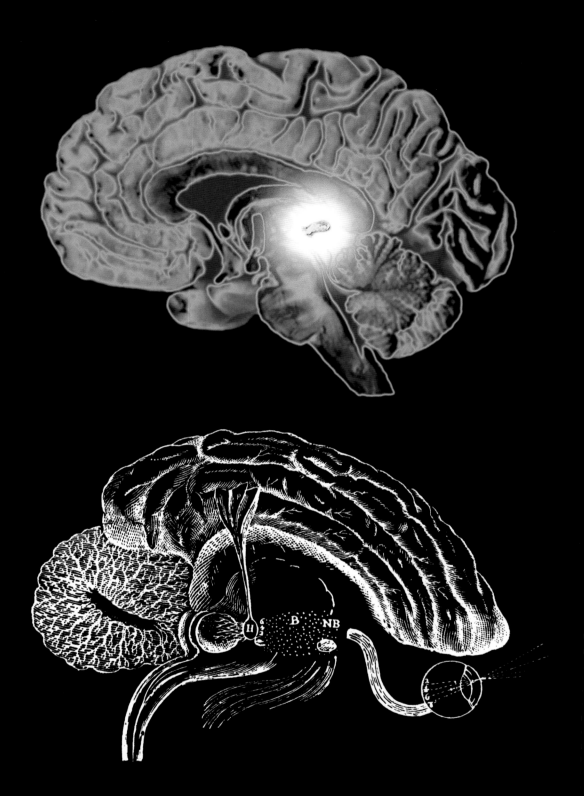

Descartes wrote that the gland was able to swivel and move around, distributing special animal spirit particles along the required nerves, which contained trap doors attached by fibres that could open or shut. He said that these particles flowed out of the ends of nerves to make the muscles swell, physically forcing them to expand while the opposite muscles contracted. According to Descarte's theory, the mind was the conscious pilot controlling the body, with the ability to override its usual actions. Subconscious activities were carried out by reactionary reflexes. Physical sensations tightened or slackened fibres in the nerves causing muscles to relax or contract. And while it had been believed that the soul provided the body with its vital heat, now a model was in place that suggested the body was independent and could survive without the soul, explaining the continuing existence of the unreasoning and ungodly kingdom of beasts. He decided that animals did not have souls and were little more than sophisticated mechanical machines.

Even though Descartes' separation of the soul from the body should have been appreciated by the Church – it removed the thorny issue of pigs in heaven – his theories were regarded as scandalous. He spent the rest of his life on the run from the dreaded Inquisition. But while his speculation about human mechanics never gained much support (anatomists knew that animals also had pineal glands), his legacy is its philosophical accompaniment. Descartes reasoned *"cogito ergo sum"* or "I think therefore I am", suggesting that the power of our consciousness and cognition makes us who we are. After the shock of discovering that the Earth was not at the centre of the Solar System, his theories gave us reason again to believe in our own importance, as the only creatures with minds. We were not an inconsequential nothing but the centre of our own Universe. The workings of the brain became the key to the secret of what makes us human.

Left: A modern computer-generated cross-section of the human brain (top) with Descartes' all-important pineal gland highlighted, compared with Descartes own interpretation (the pineal gland is marked with an "H").

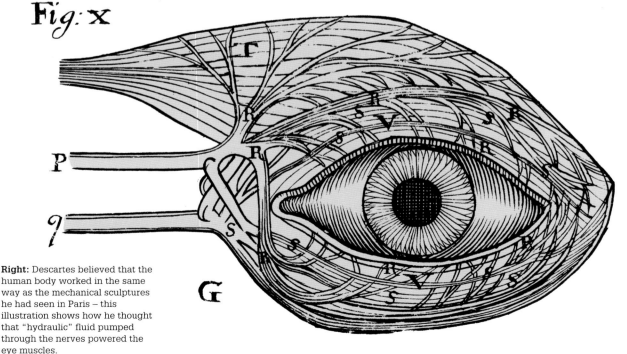

Right: Descartes believed that the human body worked in the same way as the mechanical sculptures he had seen in Paris – this illustration shows how he thought that "hydraulic" fluid pumped through the nerves powered the eye muscles.

THE BRAIN REVEALED

In the 16th century, the scholar Thomas Moore stated: "This marrow in man's head shows no more capacity for thought than a coke of suet or a bowl of curds." The physician Thomas Willis did not agree; he decided the brain was worth another look and set about proving him wrong. His cavalier attitude almost certainly stemmed from a quirk of the English Civil War. Born in 1621, he attended the village school in Great Bedwyn, Wiltshire (where he gave his lunch to the poor so often that his father worried that he was starving himself to death and made him eat at home) and was accepted at Oxford University to study medicine. The course at the time was a decade-long slog through the classical texts, including those of Galen. When war broke out in 1641, however, Oxford became a Royalist garrison and his studies were drastically curtailed. Ironically, this meant he escaped being brainwashed with Galenic theory and was able to follow his own medical interests. And these were to prove particularly perceptive.

Thomas Willis
1621–1675

A polymath, Willis was interested in all fields of anatomy and physiology. He became the richest doctor in London, one of the first specialists in diseases of the nervous system, and was buried in Westminster Abbey. But it is for his cerebral exploration that Willis is best remembered. Using the new idea that the mind, and not just God, may be responsible for determining who we are, he was the first to systematically examine every part of the brain's anatomy. Rather than seeing it as a blob of useless jelly, he tried to reconcile bodily functions with the various regions. He started by dissecting the brains of vast numbers of different animals – from earthworms to sheep – as well as human victims of a meningitis epidemic or those who were unfortunate enough to find their way to the gallows. Next, he press-ganged a fellow Royal Society member, the famous architect Christopher Wren, into creating anatomical drawings with a previously unknown level of detail. In 1664, he published

"For the first time in history someone had claimed, with proper backing, that 'who we are' was defined by our brain."

Cerebri Anatome, a seminal work that was the first to use the term "neurology", as well as the words "lobe", "hemisphere", and "peduncle" to describe parts of the brain. But of greater importance were his conclusions, which redirected our understanding of the human brain along a path that would lead towards its current interpretation. While he did not describe the neocortex, specifically, as the metaphorical mahout on the elephant, he was certain that there was a hierarchy of brain functions. He decided that there was a lower brain common to all species – referring to the cerebellum and nearby areas – but that only humans had a fully developed higher brain, which he identified as the cortex. The process of imagination he considered to take place in the corpus callosum, a strip of nerve fibres running longitudinally through the centre of the brain, linking the left and right hemispheres. The human soul, he said, was contained within the higher hemispheres. For the first time in history someone had claimed, with proper backing, that "who we are" was defined by our brain.

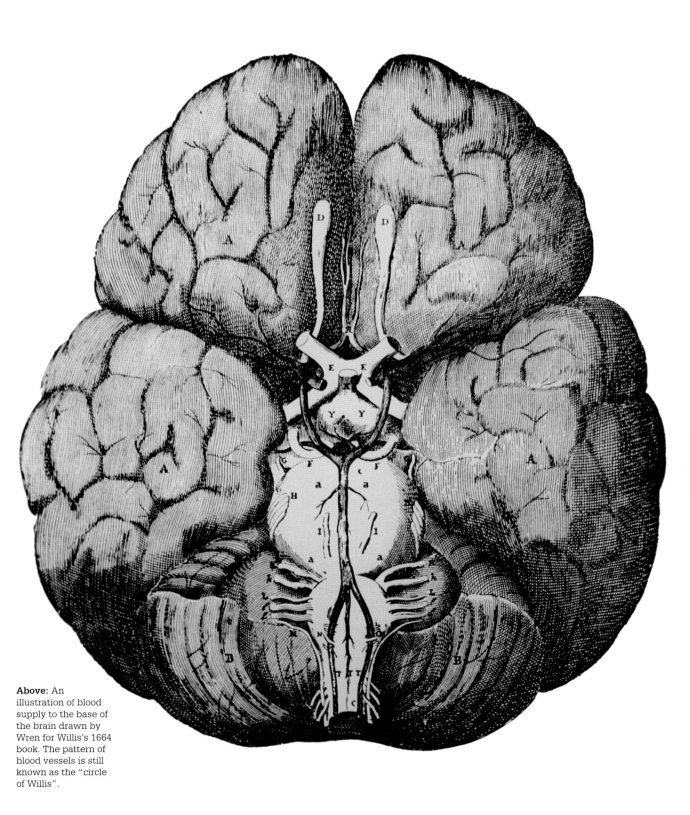

Above: An illustration of blood supply to the base of the brain drawn by Wren for Willis's 1664 book. The pattern of blood vessels is still known as the "circle of Willis".

In a nod to 17th-century pet lovers, Willis did concede that beasts also had souls, but suggested these were entirely connected to their body and would therefore perish with it at death. He also still believed in Aristotle's "animal spirits" being the particles conveyed along the nerves and even decided that these were formed in the blood vessels on the surface of the brain. As the Danish anatomist Niels Stensen said in direct criticism of Willis's physiological theories and the centuries of suppositions about how our mind controlled our body: "Animal spirits; the more subtle part of the blood; the vapour of the blood; and the juice of the nerves. These are names used by many, but they are mere words, meaning nothing." He correctly pointed out that the physiological theories of Willis and Descartes, interesting as they may have been, were little more than guesswork. No one really knew how the mind actually functioned. And nor would there be any real advances in understanding the brain's physical mechanisms until the 19th century and the discovery of animal electricity and the cell. But at least Willis had recognized where our cognition lay: enveloped deep within the cortex.

CEREBRI
ANATOME:

CUI ACCESSIT

NERVORUM DESCRIPTIO
ET USUS.

STUDIO
THOMÆ WILLIS, ex Æde Christi
Oxon. M. D. & in ista Celeberrima
Academia Naturalis Philosophiæ Pro-
fessoris Sidleiani.

LONDINI,
Typis Ja. Flesher, Impensis Jo. Martyn & Ja. Allestry
apud insigne Campanæ in Coemeterio
D. Pauli. MDCLXIV.

"Animal spirits; the more subtle part of the blood; the vapour of the blood; and the juice of the nerves. These are names used by many, but they are mere words, meaning nothing."

Left: Willis's great work on brain anatomy combined his own studies with the discoveries of several of his contemporaries.

SAVANT SYNDROME

When Dustin Hoffman portrayed autistic genius Raymond Babbitt in the film *Rain Man*, audiences struggled to believe that someone could beat the cards at a Las Vegas casino yet be too disabled to follow a conversation. However, the fictional Babbitt's incredible condition, known as Savant Syndrome, has been recognized by neuroscientists for hundreds of years. The first recorded savant, born in Derbyshire in 1707, was Jedediah Buxton. The son of a school teacher, he was deemed to have such serious learning difficulties that he was forced to work as a farm labourer, but he soon began to be recognized for his extraordinary skills at mental arithmetic. When asked to calculate the cost of 140 nails, if the first was a farthing and each subsequent one doubled in price, he instantaneously came up with the correct answer: £725,958,096,074,907, 868,531,656,993,638,851,106 2s 8d. Another 18th-century savant, Gottfried Mind, was incapable of proper communication but could perfectly recall any animal he encountered and was able to draw it from memory. Such "islands of genius" in damaged minds remain the source of much study, with scientists aiming not only to find a cure but also to unlock the secrets of memory. Current theories suggest that the condition may be caused by a faulty connection between the two hemispheres of the brain, creating an uneven balance in favour of the right-hand side – the side that handles more directly perceived and concrete thought – at the expense of the more logical left.

ENLIGHTENED MINDS

John Locke
1632–1704

Difficulties in examining the neural fabric did not stop people from questioning the basis of their existence and theorizing about what makes us, as humans, who we are. One of the most perceptive was John Locke, who paved the way to the future discipline of psychology. He studied under Willis at Oxford before taking on the role of doctor to the Earl of Shaftesbury and becoming involved in the slave trade. His "Essay Concerning Human Understanding", written while he was in exile in Holland during the Glorious Revolution that saw King James II of England overthrown, became a cornerstone of the American constitution. Indeed it was of such importance that he is rumoured to have made more from it than his contemporary John Milton received for his epic 12 book poem, *Paradise Lost*.

Locke was fascinated by Descartes's idea that "the self" consisted of our thoughts, but decided it needed a major adjustment. His explosive idea was to transform "the self" from an internal monologue to our history of conversations with the world. He said that what we are is the sum total of our life experiences, our minds a vast storeroom in which all of our memories are carefully kept until they are needed. This warehouse, he said, was not built ready stocked; our minds at birth lacked preordained ideas but entered the world in a blank slate – *tabula rasa*. Who we eventually become was determined by a continual stream of experiences. Today, the idea that our experiences of the world contribute – at the very least – to the way our mind works is so ingrained into our way of thinking about ourselves that it hardly needs stating, but in Locke's day it was a radical theory. Before Locke, it was widely believed that the soul or self was something we were born with and as the soul was thought to be immortal, it could not really change. Locke realized that if his ideas were correct then this placed much greater emphasis on the importance of early upbringing. In 1693, three years after publishing his essay, he followed it with another work: "Some Thoughts Concerning

Left: Locke's statue among the great scientists and philosophers decorating the Royal Academy in London.

Education," in which he proposed that children were not miniature, preformed adults but blank canvases whose formative years would affect their character and behaviour for life. What they learnt could literally determine whether they became good or evil. It is not surprising then that this period saw the arrival of the first children's toys designed with self-improvement in mind.

Below:
According to Locke, play and education had a crucial role in the development of a child's character and intellect.

Locke's theories had a profound effect, and not just on the field of education – they were a contributing factor in revolutions all over the world. The idea that we are all created equal became the basis of the American constitution, adopted in 1787. In France, meanwhile, cries of "Liberty, Equality, Fraternity" were given further weight by his theories; it is hard for a king to claim that he is inherently special, with a divine right to the throne, if his subjects believe his mind is simply the sum of years of pampering and privilege at their expense. The French and other European revolutions were the culmination of a century in which debates about the self and "who we are" had become increasingly important and interesting. In Britain, the new Georgian order of refinement, newspapers, and coffee houses became linked with an increasing sense of disgust about human bodies, which were now seen as comparable to animals in their pustular imperfections. In polite society at least, bodies were covered up under powders and wigs while "the mind" – our refined and entirely human possession – became an ever more elevated and treasured possession.

"Locke proposed that children were not miniature, preformed adults but blank canvases whose formative years would affect their character and behaviour for life."

AN EMOTIONAL RETURN

While humans did seem capable of behaving in a reasonable and rational manner most of the time, it was also abundantly clear that "thoughts" and "consciousness" were not all there was to being human. Our animalistic passions had to be faced before science could understand the mind; someone needed to examine the complex world of emotions. And it was the most famous natural scientist of the 19th century, Charles Darwin, who took the first significant scientific steps into the murky world of emotions. He had read an argument by the Scottish surgeon and physiologist Charles Bell claiming that specific facial muscles had been designed by God so that humans could express emotions. This went against everything Darwin believed in. So, in 1872, with the publication of his new book, *The Expression of Emotions in Man and Animals*, he set out to demonstrate that it was not only our physical characteristics that we have inherited from animals. In this vast appendix to *The Descent of Man* – his work on human origins – he attempted to show that our behaviour, our facial reactions to our mental state, and our emotions themselves are inherited traits.

Left: A page of photographic plates from Charles Darwin's work on *The Expression of Emotions in Man and Animals*.

Darwin had studied animals, such as dogs, cats, and horses, in social situations, and noted that animals not only show basic emotions such as rage, timidity, and lust but can also express more complex ones. A dog can be jealous if its master plays with another dog or even show a sense of humour as it briefly but deliberately holds back from returning a ball it knows its owner wants to throw for it. Darwin used photography to capture the expressions on animals' faces as they reacted to brief emotional occurrences and compared the photographs to those showing facial expressions of people from different countries. (Despite his publisher's concerns about effects on profit margins, seven plates of photographs comparing these expressions were used in the book, which became a best-seller). Through this study, Darwin was able to show not only that humans from every corner of the planet display the same facial reactions, but that these are fundamentally the same as those in other mammals. Applying his theory of natural selection, he suggested that, like all traits, the expressions betraying our emotions must have had biological advantages that had led to them being acquired over time. He thought that we sneer because we used to sniff at something suspicious; that we show our teeth when angry because it is an action that prepares the jaws to bite; and that we open our eyes wide in surprise as if we are dilating our pupils to see more clearly. Most importantly though, Darwin recognized that as some of these facial expressions no longer had any straightforward survival value the fact that they are the same in all humans, including

"Darwin was able to show not only that humans from every corner of the planet display the same facial reactions, but that these are fundamentally the same as those in other mammals."

TOURETTE SYNDROME

In 1825, the respected Parisian physician Jean-Marc Itard reported in the journal *Archives Générales de Médecin* the case notes of a local aristocratic lady with a most embarrassing affliction. The 26-year-old Marquise de Dampierre was plagued by a tendency in the middle of polite and respectable conversations to suddenly bark, squeal, make extraordinary movementsd and to shout out *merde* (shit) or *foutu cochon* (filthy pig). Despite her obvious embarrassment at her unconscious outbursts, Itard was unable to find a cure and it was not long before her friends deserted her. She eventually ended up in Charcot's Salpêtrière Hospital before her lonely death in 1884. However, her behaviour greatly interested the neurologist's young assistant, Georges Gilles de la Tourette, who had noted similar symptoms in other patients. The following year, he used Madame Dampierre as his primary case study in his classification of a new ailment, noting that it usually commenced at a young age with involuntary motor and vocal tics and progressed towards coprolalia – literally translated from the Greek as "talking crap". Tourette syndrome, as it became known, is still not properly understood, but illustrates the complexity that scientists often face in trying to understand who we really are.

islanders who have been isolated for tens of thousands of years, shows that we did not learn these characteristics. They must have been inherited from our common ancestors. He realized that we feel and display similar emotions as animals.

The claim that human passions are simply a continuum of animal survival patterns seemed to fit with a theory first suggested by Aristotle: we are like animals in every way apart from a higher soul – now seen as our reasoning or mind – that is able to suppress or override our base desires. This idea was very popular in the stiff and tightly buttoned Victorian age. But Darwin's theory also opened a Pandora's box. It was now impossible to scientifically consider who we are without delving deeper, going beyond our purely logical, rational, and conscious minds.

"Darwin's theory also opened a Pandora's box. It was now impossible to scientifically consider who we are without delving deeper, going beyond our purely logical, rational, and conscious minds."

Right: Another page of striking plates from Darwin's book on emotions. More than 5,000 copies were sold, making the book something of a bestseller despite its luxurious and expensive production.

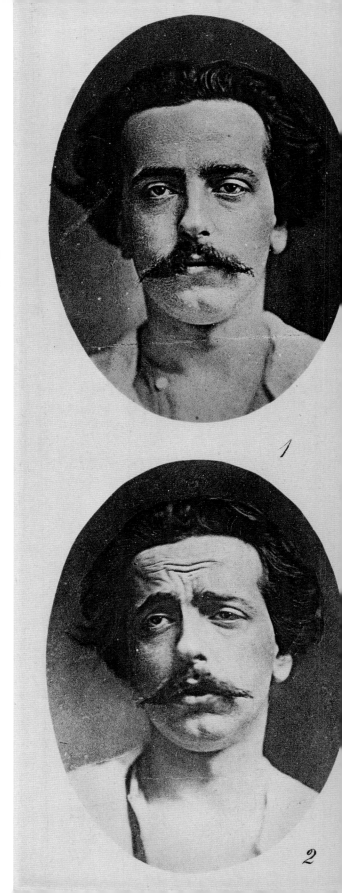

1

2

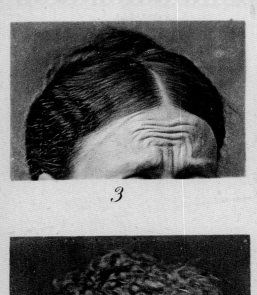

3

4

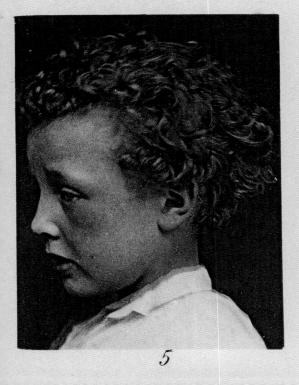

5

6

7

WHAT LIES BENEATH

If the director of a horror film was looking for inspiration for a 19th-century lunatic asylum, they would do well to read up on the Salpêtrière Hospital in Paris. Originally designed as a vast gunpowder store – one of the ingredients of which (saltpetre) gave it its name – it was eventually put to a different use as a gigantic prison for female vagrants, prostitutes, and undesirables. After the Revolution it took on the role of a mental hospital, but one in which thousands of inmates were mixed together randomly and treated little better than animals, often chained to the walls. It is hard to imagine the emotional chaos that must have reigned inside that building.

Jean-Martin Charcot
1825–1893

Having worked at Salpêtrière as a trainee doctor, in 1862, aged 37, Jean-Martin Charcot, the son of a carriage maker, accepted the role of its medecin or principal doctor. He was incredibly enthusiastic about this appointment. "This great asylum holds a population of five thousand persons, among whom are to be counted a large number who have been admitted for life as incurables; patients of all ages, affected by chronic diseases of all kinds, but particularly by diseases of the nervous system," he wrote. "The clinical types available for study are represented by numerous examples, which enables us to study a categoric disease during its entire course, so to speak, since the vacancies that occur in any specific disease are quickly filled. We are, in other words, in possession of a sort of museum of living pathology of which the resources are great." Charcot, who enjoyed the company of a pet monkey, proudly hung a sign over his office stating his opposition to animal testing – after all he had more human lab rats than he could ever need. He studied each patient, carefully noting the progress of their symptoms, and recorded the physical state of their brains and nervous systems when they died. He was also, however, a humane man. While he recognized the hospital's potential for advancing neurology, he also tried to make his patients' lives more endurable. One of his first actions was to categorize the different mental and neural disorders. Patients

Below: Established by Louis XIV in 1656, the Salpêtrière teaching hospital in Paris took its name from the old gunpowder works that once stood on the site.

with Parkinson's disease, for example, were separated from those suffering from psychotic disorders.

Charcot's recognition of the implications of the "resource" at his disposal, and his immense powers of observation (he even hired a multiple sclerosis sufferer to work as his tea lady so that he could constantly study her physical degeneration) allowed him to make numerous breakthroughs in understanding nervous disorders and mapping which parts of the brain were responsible for different physical faculties. Indeed, he was credited with turning neurology into a specialist discipline and a new chair at the Paris School of Medicine, for diseases of the nervous system, was specifically set up for him.

Although primarily a hard-nosed neurologist Charcot also developed a fascination with hypnotism and was one of the first to use it as a serious treatment. The 19th century was sometimes known as the "century of nerves" after the discovery of the importance of electricity spawned a new interest in invisible, often spiritual and indeed impossible forces. Mesmerism, magnetism, and metallurgy were all studied and tested for their potential uses in treating mental cases and Charcot experimented with all of these, but only became really enthused by hypnosis. Having decided that hysteria was a mental rather than physical malady, he felt that the intangible treatment of hypnosis might have benefits in treating it, reasoning that it could bring out different personalities much in the same way as some of his patients seemed to switch from one character to another. One of Charcot's patients was Blanche Wittmann, a Parisian socialite who developed such severe hysteria that she became terrified of the outside world. She was found to be particularly susceptible to hypnotism and her inductions into somnambulant states – in which she became calm and open to suggestion by the doctor – were often witnessed by other medical professionals. Charcot insisted that when hypnotized she did not become his automaton but was simply displaying an alternate self. In one experiment, conducted in front of a packed audience, Charcot gave Blanche a plastic knife and, informing her of their heinous sins, instructed her to stab members of the audience. Then, after the imagined killing spree, which she entered into with some gusto, she was told the room was empty and that she should undress to take a bath. Instead of obeying she had a violent hysterical attack and came out of her trance.

Although it is not clear that Charcot's work with hypnosis had any real medical benefits, it was further proof that we cannot understand ourselves simply in terms of our rational psyches; we are much more complicated and multidimensional. A young Austrian comparative neurobiologist who, before witnessing Charcot's elaborate demonstrations firsthand in 1885, had spent most of his career examining the nerves of freshwater crabs became fascinated by the idea of alternative psyches. Sigmund Freud, the man who was to become the godfather of the subconscious mind, stepped into the breach. He attempted to remodel the sense of self, which had seemed so well ordered and rational but was now rapidly descending into chaos. Freud suggested that Descartes had made a fundamental mistake when he stated that it was conscious thinking that makes us who we are and came up with an alternative claim: that we are actually controlled by our unconscious. Our base emotions, he said, like unseen puppet masters determine our thoughts and actions. Many of Freud's more detailed claims about the importance of sexuality, his interpretation of dreams, and his views of how repressed memories shape our lives are, to say the least, now open to question. But his central claim, that the unconscious controller of our lives is not the soul or God but something residing deep within the mind, refocused the question of "who we are" back to the physical properties of the brain.

Left: Charcot's student Freud took up his mentor's fascination with hysteria. However, his time at Salpêtrière convinced him to take up psychoanalysis rather than academic research into neurology.

Right: Charcot at the Salpêtrie hospital in 188 discussing a woman sufferi from hysteria. His "patient" i Blanche Wittm who was able perform on cue She later becan Charcot's assistant, and after his death went to work fe Marie Curie.

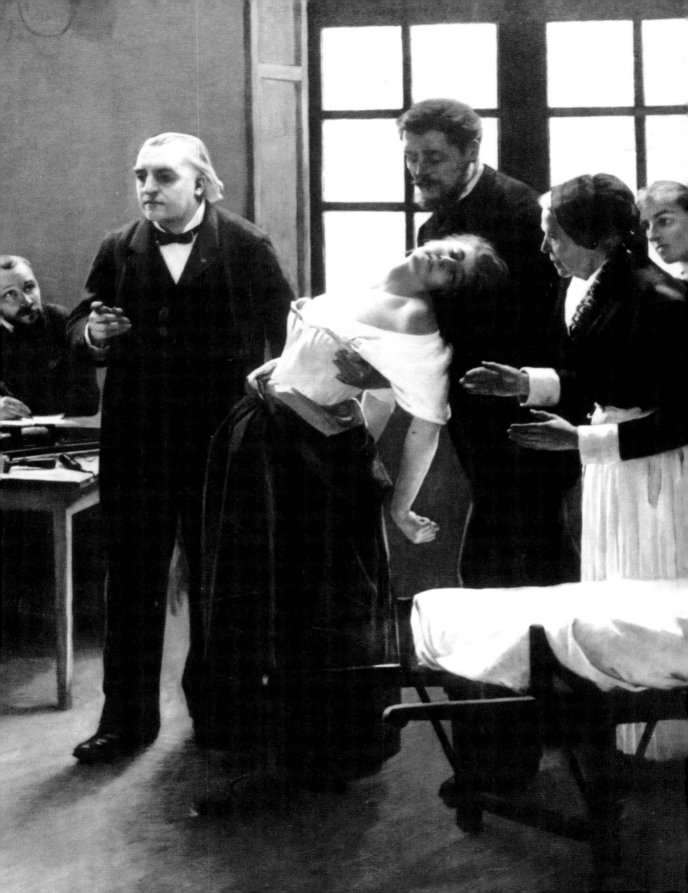

BITS OR PIECES

Up until this time, the brain had been studied on entirely conceptual or macro levels. When dissected, it was examined in terms of shape and size, either as a whole or of the various traceable regions. The composition of brain matter itself could not be usefully analysed, but it had been known for a long while that injuries to different parts of the brain seemed to have effects on different parts of the body. We have already met Thomas Willis, who divided the brain into different areas based on function. His theories were advanced in the mid 18th century by the Scandinavian Emanuel Swedenborg, who showed that different parts of the cortex controlled different muscle groups in the body. Following a series of holy visions, however, he was persuaded to give up science to become a theologian, resulting in his neurological studies – as brilliant and pioneering as they were – remaining unpublished until after his death in 1772.

Franz Joseph Gall
1758–1828

These theories of localization were eventually picked up Franz Gall, an anatomist based in Strasbourg, who decided that different areas of the cortex must be linked not only with different areas of the body, but also with aspects of a person's character. Gall was something of a rogue; he travelled with a constant retinue of strange pets and high-living Parisian women. (As he was fond of saying, "neither sin nor friends will ever leave me".) His interest in the workings of the brain was apparently piqued when he was just nine years old – upon noticing that a classmate of his who had bulging eyes was better at remembering his vocabulary lists than he was. As he became aware of other bug-eyed children with higher than average powers of recollection, he concluded that the trait must be linked to memory. Once a trained scientist he did his best to draw conclusions from studying the insides of the brain – but without success. He decided that the only way he could really begin to understand how areas of the cortex were related to mental faculties was to look at the relative sizes and shapes of people's heads, a discipline he called cranioscopy but which later became known as phrenology. He studied the skulls of over four hundred people with known mental characteristics and soon declared that he could identify the regions associated with 27 faculties ranging from sense of colour to carnivorous instinct.

Gall's ideas were so novel and outrageous that in 1801 Emperor Franz II, the last Holy Roman Emperor, banned his publications about head shapes on pain of losing his own. Yet he became one of the most influential scientists of the day, creating a belief system that was to damage the lives of people with "wrongly proportioned" heads for almost a century. Although Gall's methods were deeply flawed, his theory of localization of brain parts continued to be hotly debated throughout the 19th century. The French physiologist Jean Pierre Flourens decided that Gall was misguided; there were only three loose functional regions of the brain – the cerebellum controlling movement, the medulla, involved with vital functions, and the cerebral cortex, involved in perception and intellectual abilities. He tried to prove the latter point by successfully keeping a pigeon alive after surgically removing its cerebral hemispheres. And indeed, while it appeared blind and deaf, generally comatose and incapable of thought, it retained its vital and motor functions. There followed a frantic bid to work out whether the brain could be divided into areas or if it performed like one large interconnected mass.

Paul Broca
1824–1889

Scientists began a slew of publically unpopular cleavings of the brains of live tortoises, rabbits, and dogs. The German Friedrich Goltz kept one hound alive for 18 months after removing a large chunk of its cerebral cortex. Others opened up skulls and stimulated different areas with chemicals, scalpels, or electric shocks. Still, no consensus was reached, although the theory of localization looked increasingly correct.

Finally, in 1861, Paul Broca proved that the frontal lobe was responsible for speech, making it the first area of the cortex to be almost universally accepted as relating to a specific function. Meanwhile, David Ferrier, a neurologist from Aberdeen, had decided that a compromise was needed between those in favour of a holist theory and those who believed in localization. By experimenting with living dogs and monkeys he came to the following conclusion: "From the complexity of mental phenomena and the participation in them of both motor and sensory substrata, any system of localization of mental faculties which does not take both factors into account must be radically false." He realized that while there was no specific area of the brain responsible for intellect, other narrower functions may have their own regions. As a reward for this pearl of wisdom he was taken to court by animal rights campaigners disgusted by his countless vivisections.

By the end of the Victorian age, neurology and the workings of the mind had become frustratingly unclear. So much had been discovered in every other area of science, but in relation to the brain, there were still far more questions than there were answers. The instrument that could finally unlock some of its secrets – the high-power microscope – had finally arrived, but a further development was needed before this could be put to use.

PSYCHOSURGERY

Phineas Gage was a construction foreman working on the Rutland to Burlington railroad in Vermont in 1848. One day, as he was packing a large quantity of high explosive into a deep but narrow hole in a rock face, a rogue spark ignited. As the smoke cleared, Gage stumbled down the hillside having had his tamping iron – a three foot long metal rod an inch and a half in diameter – blown straight through the frontal lobes of his brain. Unbelievably Gage survived. His character, however, changed so dramatically that his friends and family no longer recognized him as the same person. Gage, once responsible and popular, was now a rude and distant oaf. His widely reported recovery and change of character led some scientists to believe that the brain could and should be operated on to cure psychological problems, particularly after suspicions arose that severing the frontal lobes from the rest of the brain could make people less anxious or emotional. This idea was taken up in earnest in the mid 20th century by Walter Freeman, a laboratory director for a Washington mental hospital. Anxious to quieten his miserable and delirious charges, he developed the "transorbital lobotomy". He used an ordinary ice pick, which he would plunge through the patient's tear duct before waggling it around, to slice though the tissue of the brain. Despite regularly leaving his patients worse off than Phineas Gage, he achieved enough of a reputation to carry out over 3,000 such unregulated "psychosurgical" operations between 1936 and 1967.

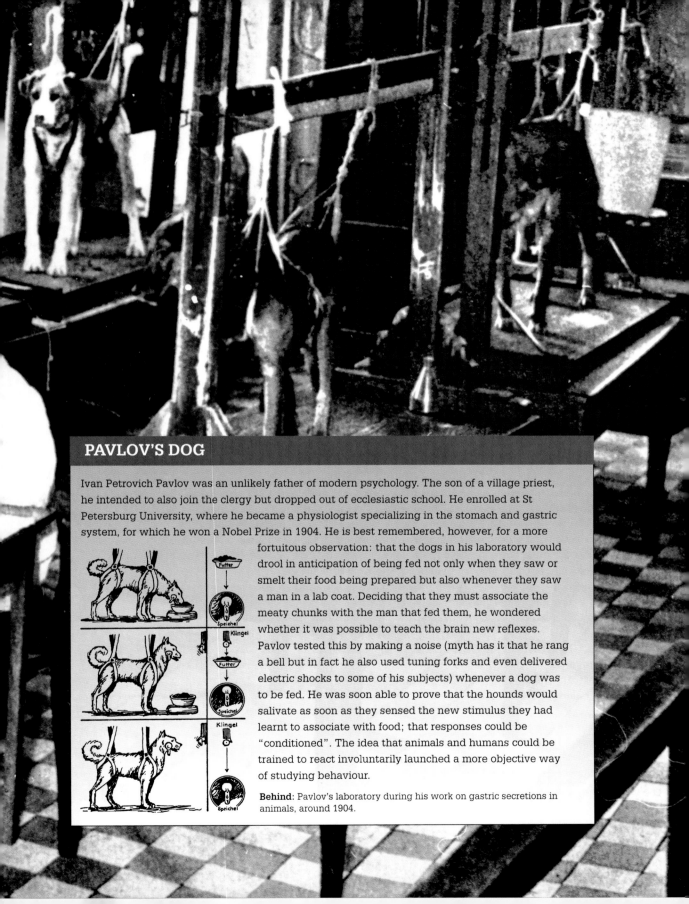

PAVLOV'S DOG

Ivan Petrovich Pavlov was an unlikely father of modern psychology. The son of a village priest, he intended to also join the clergy but dropped out of ecclesiastic school. He enrolled at St Petersburg University, where he became a physiologist specializing in the stomach and gastric system, for which he won a Nobel Prize in 1904. He is best remembered, however, for a more fortuitous observation: that the dogs in his laboratory would drool in anticipation of being fed not only when they saw or smelt their food being prepared but also whenever they saw a man in a lab coat. Deciding that they must associate the meaty chunks with the man that fed them, he wondered whether it was possible to teach the brain new reflexes. Pavlov tested this by making a noise (myth has it that he rang a bell but in fact he also used tuning forks and even delivered electric shocks to some of his subjects) whenever a dog was to be fed. He was soon able to prove that the hounds would salivate as soon as they sensed the new stimulus they had learnt to associate with food; that responses could be "conditioned". The idea that animals and humans could be trained to react involuntarily launched a more objective way of studying behaviour.

Behind: Pavlov's laboratory during his work on gastric secretions in animals, around 1904.

Even in the Behaviourist camp, however, not everyone accepted Watson's simplified view of humanity. Burrhus Frederic Skinner, a hobbyist inventor and Harvard psychologist, realized that people, no matter how young, could not be described as input-output machines; emotions as well as reasoning would always affect behaviour. He decided that in order to study the brain objectively another control had to be introduced. His solution was the "operant conditioning chamber", also known as the Skinner Box. These, versions of which are still used in psychological studies today, were special cages in which an animal, usually a rat or a pigeon, was placed. The boxes contained levers that would deliver some sort of tasty morsel as a reward, as well as areas that could deliver electric shocks or punishments. Skinner realized that, in this isolated environment, if an animal responded in one way to an event the probability increased that it would behave the same way again. Behaviour could be scientifically studied and possibly even predicted.

Left: As the originator of behaviour studies, B.F. Skinner (shown here with an experimental subject in a "Skinner Box"), has been cited as the most influential psychologist of the 20th century.

Of course, in the real world with its infinite choices, experiences, and sensations, we know that the ability to successfully make such predictions still eludes us. Despite the extraordinary power of computers, which sprung from the launch pad of World War II and the scrabble to crack the German Enigma code, we still do not know enough about human logic to be able to replicate it. The complexity of our reasoning is still too great to truly understand our "self". Alan Turing, often regarded as the father of artificial intelligence and one of the scientists who worked at Bletchley Park, where codes were cracked in Britain during the war, questioned whether machines can think. So far, the answer remains no. Even if we are behaviour-response machines, we are still much too complicated to replicate.

ILLUSIONS AND MAGIC

We may think that we can believe our own eyes, but our perceptions are actually influenced by our imagination. The pictures formed in our minds may not entirely reflect reality but instead be what we expect to see based on past experiences and associations. This is why optical illusions work. A simple example of this is a drawing called "Message of Love from the Dolphins" by Sandro Del-Prete – adults almost always see the image as a naked couple locked in an intimate embrace while children see a more innocent image of nine dolphins. Communication between our mind and our eyes is seldom perfect, no matter how good our vision is. If we stare at an escalator moving downwards for long enough and then glance across at a stationary one it will appear to be moving upwards. This is because cells in the brain that respond specifically to downwards motion become tired and unresponsive, meaning that those responsible for upward motion will have a greater and distorting effect. Differences between what our brain is likely to perceive and what actually happens were recognized and exploited by magicians long before neurologists began to understand the principles of perception. For example, during a trick we will always look out for sneaky hand movement but assume that a pack of cards in their cellophane wrapper, complete with jokers, will be the full deck. Simple as it sounds, magicians know that our preconceived ideas about packs of cards will lead us down the wrong path. It is worth remembering that a magician's deck seldom has the correct amount of cards, even if unopened.

AND BEYOND...

Just as the telescope, in the 17th century, extended human understanding of the Universe, so modern brain-scanning techniques, like functional magnetic resonance imaging (fMRI), are opening the door to a new wealth of information about the workings of the brain. We now know, for example, that we have neurons so specific that they appear only to be interested in one thing. Scientists have found neurons that only become active when a person thinks about the actress Jennifer Aniston; Bill Clinton and Halle Berry also have their own neurons. And as early as 1960 a neuron was identified that is only activated when a person thinks about their grandmother. Brain scanners can also reveal the areas of the brain that are active when a simple choice is being made, such as which button to press out of a choice of two. This has allowed experimenters to predict their subject's decision several seconds before the person being analysed is even aware that they have made that choice.

While we have become increasingly aware of how complex our brains are, however, we have also become aware that our brains, like ourselves, are prone to taking short cuts. Most people think of vision, for example, as being like a video camera, faithfully recording what is out there. In fact it is nothing like that. The brain tends to create its own reality and ignore most of what it sees. In one test, psychology students on their way to an interview met a receptionist sitting behind a desk. They were then briefly distracted by someone shouting their name, during which time the receptionist swapped with someone completely different. When the students turned back to the receptionist, none of them noticed they were now talking to someone completely different. Scientists searching for the cause of road accidents have discovered even more disturbing implications of this corner cutting. If you are driving down an empty street and glance away, when you look back the brain assumes it is looking at much the same image it was looking at a second before. There is often a significant time delay before it registers that the scene has changed and that there is now someone standing in the middle of the road. This phenomenon is known as change blindness. The reason it happens is that the extraordinary quantity of information that we are constantly receiving through our senses cannot all be processed instantaneously. The brain is forced to take short cuts, to make assumptions.

"Modern brain-scanning techniques, like functional magnetic resonance imaging (fMRI), are opening the door to a new wealth of information about the workings of the brain."

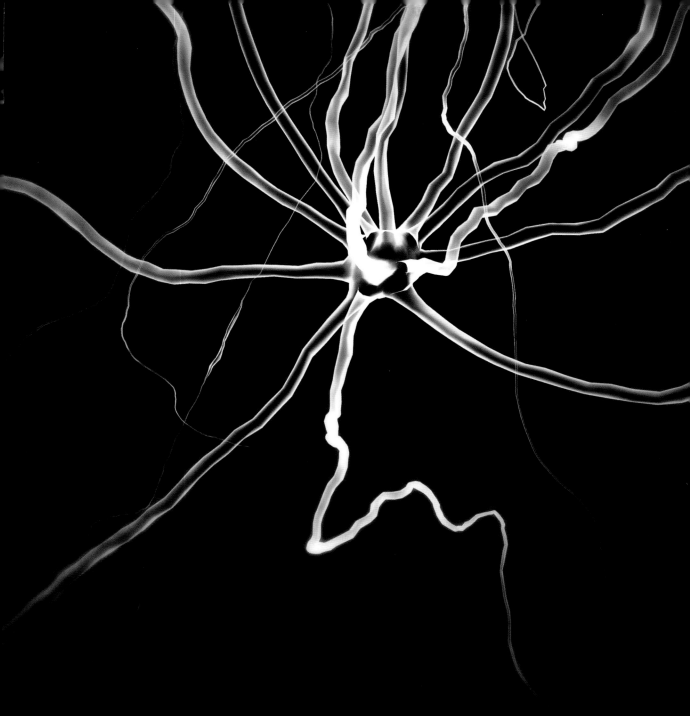

Above: Microscopic view of a neuron (signal-processing nerve cell) and its network of axons or nerve fibres – long extensions used for transmission of signals between neurons.

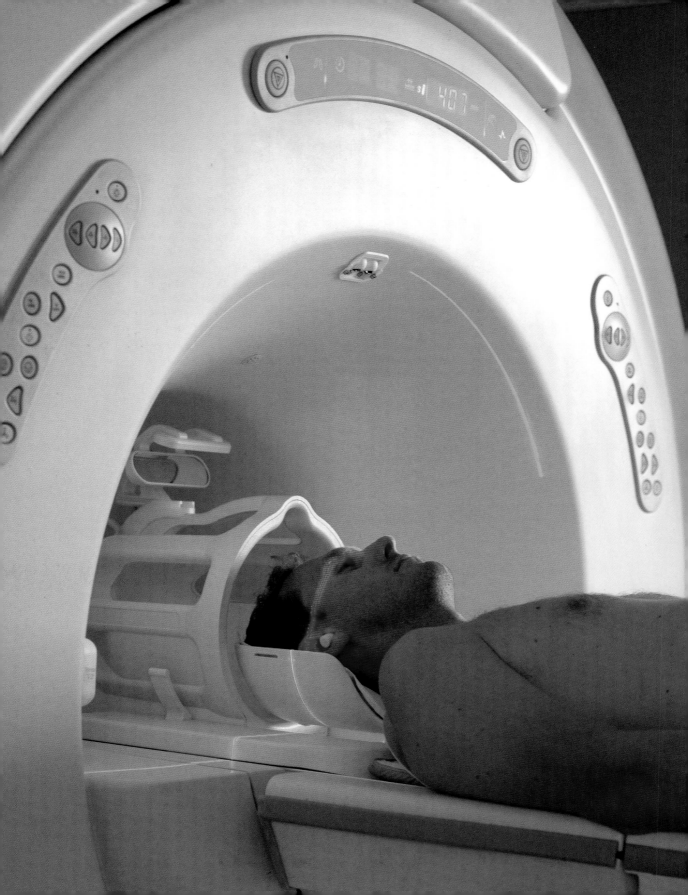

The brains acts, in a way, like a miniature scientist in residence. We unconsciously gather data, make assumptions (hypothesize), test these assumptions, and modify our beliefs accordingly. The scientific method is constantly being applied in all of our heads. It may not feel like it when you sit in a chemistry lecture, but science is, in some ways, a very natural activity. Clearly most animals use "science" in the sense that they make assumptions about their environment, which they then test out, albeit at an unconscious level. Yet we are the only creatures that reflect on what we do; we go beyond our immediate experiences of the world. This is something that seems to be uniquely human – the ability to step beyond what our senses are telling us, to peel back the veil, to see behind the illusion. Often we get it wrong, but the great thing about the scientific method, more so than any other belief system, is that it allows us to learn from our mistakes. We know that whatever it may feel like, the Earth is flying through space and around the Sun, rather than vice versa. We know that though the table top may feel solid, it is mainly empty space. We know that though the world about us feels, in our limited time scale, to be unchanging and permanent, it is in fact constantly changing. And we know, through DNA, that though we may feel chosen, special, we humans are closely related to every other creature on Earth.

Left: Modern medical imaging scanners, and the ability to see how the brain functions while subjects are not only alive, but conscious, are creating a new revolution in our studies of the human mind.

Right: Magnetic resonance imaging (MRI) maps soft tissues in the body through the presence of water molecules. It can also track oxygen levels in the bloodstream, revealing active areas of the brain through their increased demand for oxygen.

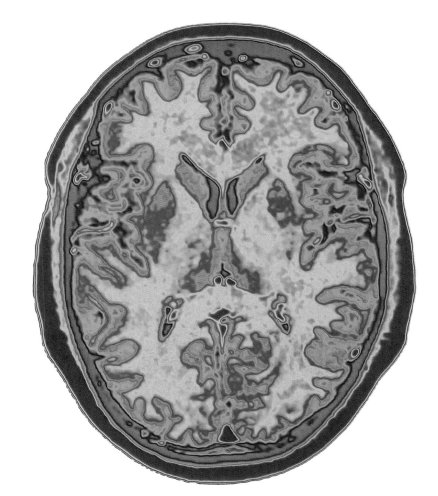

Connections – Mind

ISLAMIC SCIENCE

AGE OF DISCOVERY

Hippocrates
c460 – c370 BC

Plato
c428 – c348 BC

Aristotle
c384BC – c322 BC

René Descartes
1596 – 1650

Thomas Willis
1621 – 1675

RENAISSANCE

REFORMATION

CLASSICAL GREECE

ROMAN EMPIRE

MIDDLE AGES

^ *Trepanning*

^ *Egyptian manuscript describing effects of brain injuries*

We have been trying to understand human identity and motivations for thousands of years. Agreement that this search should focus on the brain is surprisingly recent, however. In ancient Greece, Hippocrates, like the early Eygptians before him, deduced from the effects of injuries to the brain that it had some influence on controlling the body, a belief possibly reflected in the practice of trepanning. Plato built further on his idea, but Aristotle relocated the rational soul to the heart, where it stayed for a thousand years – adopted and policed by the Church.

The anatomical observations of Andreas Vesalius and later René Descartes put the search back on track, the latter developing the theory of dualism – that the mind was separate to the body, "piloting" it from a cockpit in the pineal

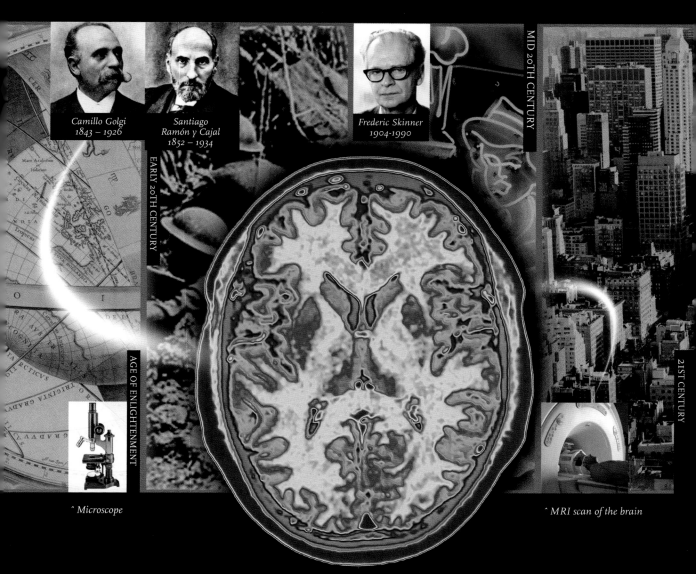

Camillo Golgi
1843 – 1926

Santiago
Ramón y Cajal
1852 – 1934

Frederic Skinner
1904-1990

MID 20TH CENTURY

EARLY 20TH CENTURY

AGE OF ENLIGHTENMENT

21ST CENTURY

^ Microscope

^ MRI scan of the brain

gland in the brain. Shortly after, Thomas Willis became the first to systematically examine the anatomy of the brain, and to definitively state that it determined who we are.

The search for the self was far from over though. Investigations continued – into how our minds differed from other creatures, into how we could be conditioned and controlled (famously by Pavlov and Skinner), and into how we could be cured of mental diseases. Throughout, the ever-improving ability to understand the physical structure and processes of the brain was crucial – early microscopes enabled Camillo Golgi and Santiago Ramón y Cajal to reveal the detailed structure of the brain and the existence of neurons. Today, with techniques such as MRI scanning, we are even able to see the brain in action.

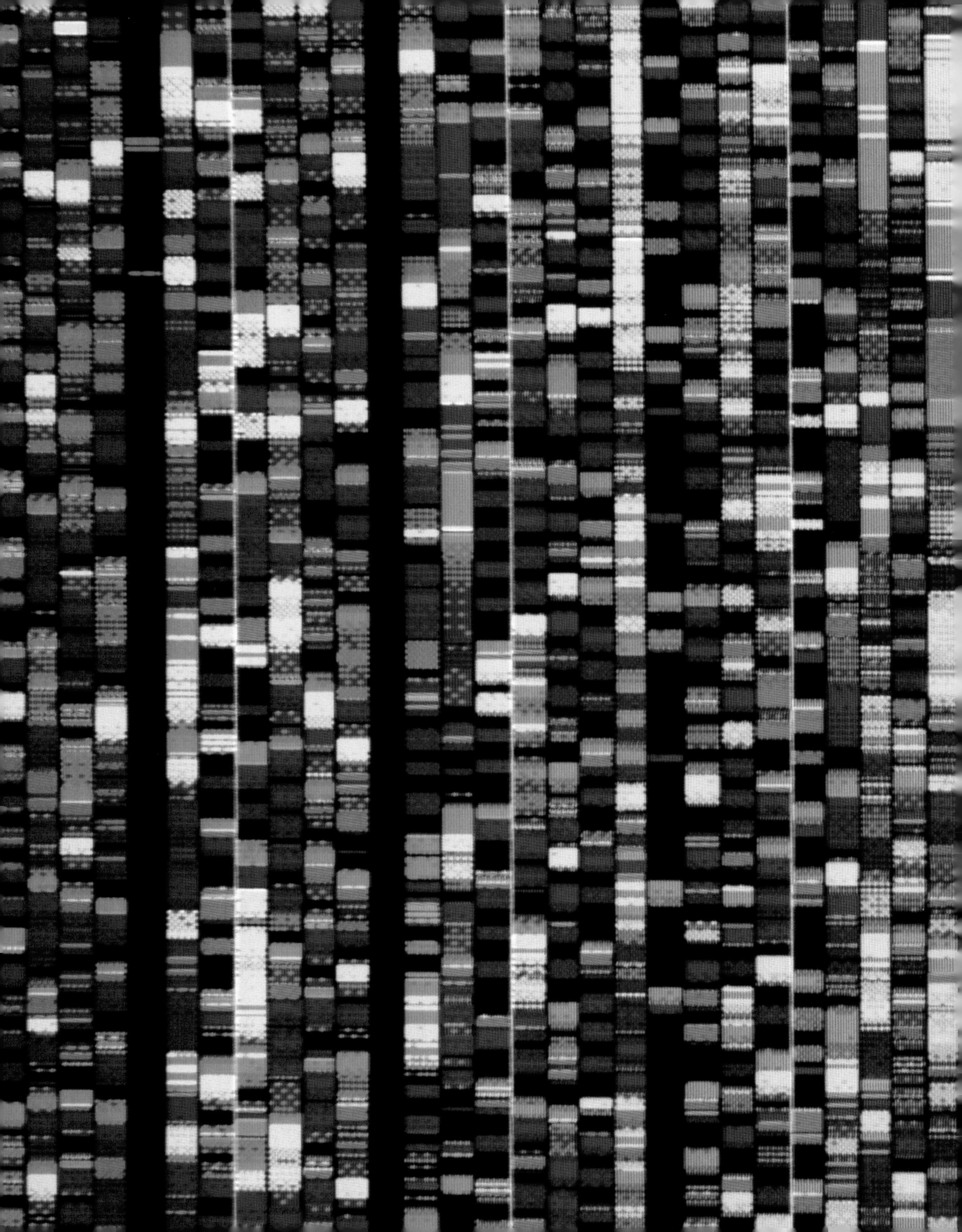

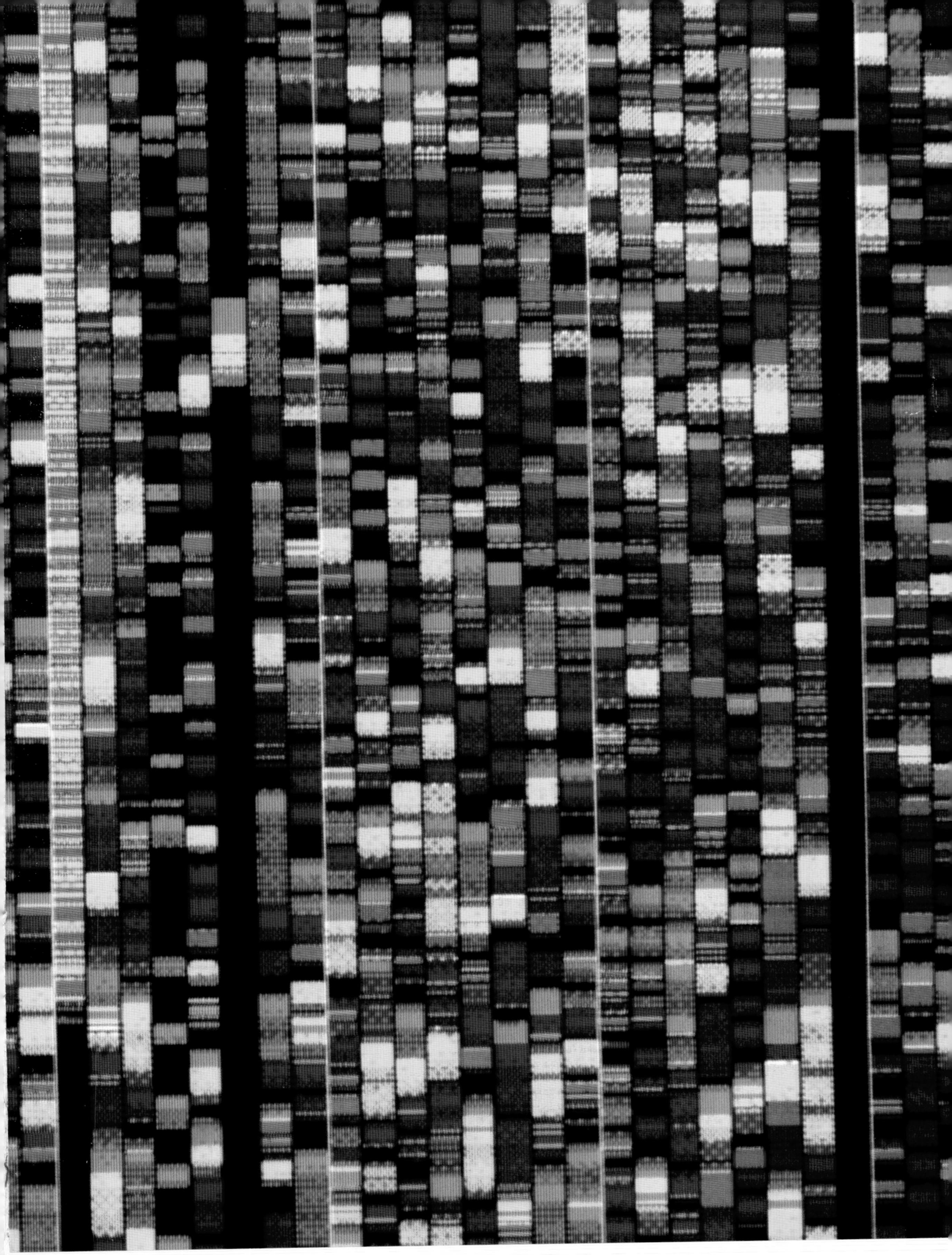

PICTURE CREDITS

Travelogue 205 left, 205 right; /A. Barrington Brown 222; /David Becker 264; /Juergen Berger 209; /George Bernard 11, 111, 114 right, 143 top centre, 144; /Paul Biddle & Tim Malyon 189 bottom; /Maximilien Brice, CERN 56; /Dr Jeremy Burgess 191 left, 242; /Caltech Archives 34; /CCI Archives 190 right, 228 top centre right; /J-L Charmet 168 top; /Russell Croman 152; /Custom Medical Stock Photo 213 top; /Dopamine 206; /John Durham 210; /Bernhard Edmaier 123; /Emilio Segre Visual Archives/ American Institute of Physics 55 top centre right, 96, 101 top centre left; /Equinox Graphics 134; /Prof. Peter Fowler 92, 101 top centre right; /Simon Fraser 200; /General Research Division/New York Public Library 118; /Steve Gschmeissner 226; /Tony Hallas 52; /Roger Harris 240; /Adam Hart-Davis 208 top right, 208 left, 229 top left; /Gary Hincks 136; /Keith Kent 155; /Gavin Kingcome 110 top; /James King-Holmes 2; /Mehau Kulyk 188, 190 left, 249; /Andrew Lambert Photography 80; /Patrick Landmann 135; /Library of Congress 146 top, 167, 172 top, 178; /Library of Congress/New York Public Library 166 bottom; /Living Art Enterprises, LLC 91; /Medical RF.com 270; /Astrid & Hanns-Frieder Michler 91 bottom; /Mid-Manhattan Picture Collection/Glass/New York Public Library 169; /Cordelia Molloy 59 bottom, 173, 179 bottom, 185 bottom centre left, 185 bottom centre right; /NASA 6, 8 top, 39, 48 centre; /NASA/ESA/STSCI/L. Sromovsky, UW-Madison 48 left; /National Library of Medicine 258; /NOAA 33; /Omikron 277; /David Parker 99, 276–277; /Pekka Parviainen 28; /Pasieka 72, 216, 229 centre, 244 top; /Photo Researchers 179 top, 185 top right, 208 bottom right, 228 bottom right, 229 top right, 274 top centre left; /Physics Today Collection/American Institute of Physics 90; /Philippe Plailly/Eurelios 140; /Maria Platt-Evans 108 top, 142 top right, 158 top, 185 top centre left; /Radiation Protection Division/Health Protection Agency 180; /Detlev Van Ravenswaay 19 bottom, 26 bottom left, 54 top left, 55 bottom left; /Royal Astronomical Society 21, 26 centre, 45, 55 background picture 2, 100 background picture 8, 142 background centre, 184 background centre , 229 background picture 1, 274 background centre; /Friedrich Saurer 32; /Jon Stokes 260 left, 275 bottom left; /Science, Industry & Business Library/New York Public Library 168 bottom; /Science Source 59 top, 64 top, 100 top left, 204, 220,

229 top centre right, 260 right; /Sovereign, ISM 230; /Sinclair Stammers 110 bottom; /St. Mary's Hospital Medical School 14; /Mark Sykes 64 bottom; /Sheila Terry 25, 36, 44, 54 top right, 55 top left, 55 top centre left, 78 bottom, 79, 79 bottom, 100 top centre, 100 top right, 116, 119, 129, 133, 143 top centre left, 171, 174 top, 182, 184 left, 185 top centre, 185 top centre right, 188 bottom, 188 top, 189 top, 194, 195 top, 197, 228 top left, 228 top centre left, 229 top centre left, 235, 237 bottom, 242 right, 250 right, 274 top centre, 274 top centre right, 274 bottom left; /G. Tomsich 193 bottom; /Michael W. Tweedie 227; /US Department of Energy 101 bottom right; /US Library of Congress 51, 55 top right; /US National Library of Medicine 101 background picture 1, 143 background picture 1, 185 background picture 1, 229 background picture 2, 261 Bottom, 275 background picture 1; /Charles D. Winters 76; /Ed Young 98.
The Natural History Museum, London 104 bottom, 105, 114 left, 142 top left, 142 bottom left.
TopFoto 169 left; /Fortean 24; /The Granger Collection 38, 93.
Wellcome Library, London 237 top, 248, 253, 252, 254.
De Beer Collection, Special Collections, University of Otago, Dunedin, New Zealand 126, 143 bottom left.
Richard Wheeler, Sir William Dunn School of Pathology, University of Oxford 224–225.

ACKNOWLEDGEMENTS

The television series on which this book is based is presented by Michael and executive produced by John, for whom a television history of science has long been a goal. The concept of the series has its origins in the development team of BBC Science, particularly Ros Homan who set out the framing concept of the six great questions. The programmes themselves were shaped and crafted under the inspiring editorial leadership of series producer Aidan Laverty. The research team of Alice Jones, Naomi Law, Liz Vancura and Will Ellerby took on what is effectively a history of everything to search out compelling stories, which were brought to life by directors Jeremy Turner, Nat Sharman, Peter Oxley, Nicola Cook, Giles Harrison and Nigel Walk, while the sprawling production process was managed with true dedication by Maria Caramelo and Sarah Forster, under the ever-wise guidance of production manager Giselle Corbett.

Countless experts and contributors gave time and effort to the preparation of the television series, but none more so than its three highly-committed academic advisers, Pietro Corsi, Jim Endersby and Patricia Fara, whose expertise, wisdom and insight guided the production team's path through the programme scripts.

For the book, we are indebted to the commissioning efforts of Peter Taylor, the editorial guidance of Georgina Atsiaris and the design talents of Pene Parker, Yasia Williams-Leedham and Mark Kan. Special credit goes to Hayley Birch for her skills in smoothing the rough edges of our draft chapters, for integrating our often differing literary styles and for keeping a weather eye open for embarrassing mistakes - although we retain responsibility for any errors that remain. A special thanks also from Michael to Ewan Fletcher for his help – and from John to DM Lawrence for her unflagging assistance in managing the day.

Finally, it is simply not possible to undertake a challenging project like this without the continued support of friends and family. Michael could not do what he does without Clare; and John has been humoured by Ewa, Christopher and Toby over the course of a difficult summer and autumn. We hope they all feel it was worth it.